THE END OF DIABETES

THE END OF DIABETES

The Eat to Live Plan to Prevent and Reverse Diabetes

Joel Fuhrman, MD

HarperOne
An Imprint of HarperCollinsPublishers

HarperOne

HarperCollins books may be purchased for educational, business, or sales promotional use. For information, please e-mail the Special Markets Department at SPsales@harpercollins.com.

HarperCollins website: http://www.harpercollins.com

HarperCollins®, 📖®, and HarperOne™ are trademarks of HarperCollins Publishers.

FIRST HARPERCOLLINS PAPERBACK EDITION PUBLISHED IN 2013

Interior Design by Laura Lind Design

Library of Congress Cataloging-in-Publication Data is available upon request.

ISBN 978–0–06–221998–5

18 LSC(C) 20 19 18 17 16

In memory of Daniel Boller, a wonderful young man,
taken by the vicious consequences of diabetes

A CAUTION TO THE READER

If you are taking any medication, do not make dietary changes without the assistance of a physician, as medication adjustment will be necessary to prevent excessive lowering of the blood sugar level (hypoglycemia). Hypoglycemia from using too much medication can be dangerous.

Because this diabetic reversal program is so effective, it is even more important to consult with a knowledgeable physician who is familiar with the medication reduction needed as a result of aggressive dietary modifications. Do not underestimate how effective this program is because without medication reductions, a serious hypoglycemic reaction could occur from taking too much medication.

Many physicians, not realizing how effective this diet style is, may be hesitant to taper medications sufficiently. Make sure you warn your physician about this and follow your blood sugar more carefully the first few weeks after beginning this plan. If you are also on medications for high blood pressure, this nutritional advice may also lower your blood pressure too much, so be sure to watch that and discuss any changes with your physician as well.

I will discuss medications in detail and offer guidance for their reduction in this book. You must realize, however, that a book cannot take the place of individualized council from the physician who knows your medical condition. It is your responsibility to work with the physician of your choice to assure your blood sugar and blood pressure readings are not too high or too low.

Note: The cases in this book are all real, but the names have been modified for privacy purposes.

Contents

THE END OF DIABETES

A Letter of Hope

DEAR FRIEND,

Congratulations. You have taken the first step in freeing yourself from the life-threatening disease known as diabetes.

As you may have read or heard, nearly 26 million Americans (11.3 percent of adults) now have diabetes, according to the Centers for Disease Control and Prevention new estimates of diagnosed and undiagnosed diabetes. Nearly 80 million (35 percent of adults) qualify as having prediabetes. If the trend continues, one in three American adults could have diabetes by 2050.

Diabetes is widespread, and we can no longer take a passive approach to getting our health back. This book is designed for people who want to take aggressive action in their battle to lose a dramatic amount of weight and reverse diabetes, high blood pressure, and heart disease. You can seize control of your health. It is in your hands. Together, we can start right now.

This program has been tested by thousands of individuals, and the extraordinary results have been documented in medical studies. It is possible to prevent and recover completely from type 2 (adult-onset) diabetes.

As a diabetic you probably have a plan to keep on top of your condition with glucose monitoring, HbA1C measurements, regular physician visits, and medication adjustments. These standard and accepted practices to maintain control of your blood glucose are seen as essential to your health. Unfortunately, this is all wrong. Your life and these treatments are focused on controlling your blood sugar instead of learning how to rid yourself of diabetes. Even with adequate glucose control, if you remain diabetic, the illness will age you prematurely and shorten your life. What's more, when you focus only on the numbers instead of removing the causes of diabetes, it could actually worsen your diabetes in the long run.

The majority of medications used to lower blood sugar place stress on your already failing pancreas. The probability of your diabetes getting worse under conventional medical care is especially likely since medications used to control blood sugar, such as sulfonylureas and insulin, also cause weight gain. The dangerous combination of pushing the pancreas to produce insulin and gaining more weight with medication actually results in the need for additional medication as you become increasingly diabetic. This common and yet failed approach shortens life span and increases the risk of heart attacks.

The number of people with type 2 diabetes is rapidly increasing, having tripled in America over the last thirty years. The main reason for this is openly recognized: America's expanding waistline. Yet physicians, dieticians, and even the American Diabetes Association (ADA) have all but given up on promoting weight loss as the primary treatment for diabetes. Medication is the accepted treatment—even though it is often the medication itself causing more weight gain, worsening symptoms, and making individuals more diabetic. This creates a vicious cycle: as a person becomes more diabetic, more medications are needed, the doses keep going up and up, and the person become more diabetic. It is a misguided approach to our health. Most diabetics would be better off if these medications were never invented because maybe then they would have been forced to change their lifestyle and eating habits. Reversing and preventing diabetes on an individual and national level does not require a prescription. It requires a change in the way we eat.

The medical community has given up on weight loss as an avenue to help diabetics mostly because traditional diets don't work. But even if you have failed on one diet after another in the past, please don't give up. The diet plan in these pages *does* work. You will see radical improvements in your health. You are the owner and operator of your body. You *can* reverse and even eliminate your diabetes with the life-saving nutritional information in this book. The nutritional plan I have used for over twenty years on more than ten thousand patients is based on a central idea:

Your Health Future (H) = Nutrients (N) / Calories (C)

My approach is radically different from other methods and is proven to work. I will show you how your body can heal itself when you give it the necessary tools. The fact is, your body is designed for wellness. Give it the right biochemical environment for healing and it becomes a miraculous self-healing machine. My approach is based on a scientific formula that determines life span and health. This formula, known as H=N/C, means your health is determined by the nutrient-per-calorie density of your diet. When you eat more foods that have a high-nutrient density and fewer foods with a low-nutrient density, your health will dramatically improve and your diabetes will melt away.

When you eat mostly high-nutrient foods, the body ages slower and is armed to prevent and reverse many common illnesses. The natural self-healing and self-repairing ability that is hibernating in your body wakes up and takes over, and diseases disappear. A nutrient-rich menu of green vegetables, berries, beans, mushrooms, onions, seeds, and other natural foods is the key to achieving optimal weight and health.

Contrary to popular speculation, the many diseases that plague all people and threaten our lives are not an inevitable consequence of aging. We are not the victims of poor genetics. We do not need a steady supply of pills for the rest of our lives. We have come to believe that our excess, disease-causing body fat is normal, acceptable, and too difficult to take off. Drugs are not the solution to the weight, diabetes, or other problems that seem to come with aging.

Knowledge leads to power. Learning how the foods you eat affect your health and well-being gives you the power to become healthy, live longer, and feel better every single day. People who use my program are amazed by the results. When you eat sufficient micronutrients and fiber with a high-nutrient diet, it suppresses food cravings. Amazingly, you begin to naturally crave fewer calories. This puts an end to over-eating. If you are overweight, this approach will rapidly create weight loss until your body finds its natural, healthy weight. For most people, the weight loss obtained through this diet rivals that of gastric bypass surgery but without the risk.

I know you're thinking, *Will I be hungry all the time? And will the food be good?* Here's the great news: healthy food should be and can be easy to prepare and delicious. I have traveled the world and have worked with celebrated chefs to come up with recipes and meal plans that are filling, mind-blowingly delicious, and good for you. No kitchen expertise is required, as these recipes are for everyone. As you follow this diet, I promise that it will soon become the way you prefer to eat. So many of my patients who started Eat to Live have changed the way they eat forever. The food tastes good, and they feel good. The truth is that once people understand the fundamentals and amazing rewards of healthy eating, we never go back to our old habits. This approach is priceless because it is lifesaving.

> Your health is dependent on the amount
> of nutrients in your diet.

I call a diet that is rich in micronutrients a *nutritarian diet*. In other words, the more *nutrient dense* your diet, the healthier you become. It sounds so simple, and it is. When you eat a diet rich in healthy, natural foods from the earth, you give your body the nutrients it needs to heal and protect itself. Diabetes is a food-created disorder, and the right food choices can rid you of this life-shortening disease and its associated medical complications.

Of the more than ten thousand patients I have counseled, many who came to me sick, overweight, and suffering from a health crisis, most have found the solution they sought for so long. They recovered and returned to health without drugs. The number one recommendation I make for all of my patients, regardless of their condition, is to overhaul their diet. I have helped thousands of people with type 2 diabetes to reduce and eliminate their disease with nutrient-dense food. A large majority of them have become nondiabetic. The results from applying this approach have been documented. In fact, I believe that my nutritarian diet, often called Eat to Live, is the most effective program for diabetes ever studied and will continue to prove to be so as it is more widely implemented and larger research studies are performed in the future.

In a case series published in the *Open Journal of Preventive Medicine,* 90 percent of participants were able to eliminate or reduce their medication by 75 percent, and the average hemoglobin A1C dropped from 8.2 to 5.8.[1] Hemoglobin A1C is a measurement of average glucose levels over a three-month period. A level lower than 6 is considered nondiabetic, or normal, and above 8 is considered poorly controlled. The participants also saw their systolic blood pressure drop from an average of 148 to 121 while medications were withdrawn. These dramatically positive results are enabling larger and more long-term studies to begin.

Of course, no dietary approach to diabetes will succeed without attention to other risk factors—the main ones being a sedentary lifestyle, smoking, and lack of sleep. The road to wellness involves making a commitment to a healthy lifestyle. Exercise is also critical. The good news is, the healthier you eat and the better you feel, the more you will want to exercise and keep your body in the best possible shape.

Yes, diabetes is a very serious disease. It can cause a host of problems such as heart disease, kidney damage, and vision loss, problems that can shorten your life and lessen the quality of your years on earth. *But it doesn't have to.* The answer is simple: eat a nutritarian diet and exercise daily. It may not always be easy, but the effort delivers life-saving results.

I urge you to take the plunge and carefully follow this program. I know firsthand that it can change the course of your health and life forever. Join us and turn the page of your health history. Let's create a new story to tell of health, vitality, and long life. It's time for a firm commitment to getting in the best shape of your life. Thousands of people have already embraced this message and are creating a health revolution. We're thrilled to have you along for this exciting and transformative journey.

—Joel Fuhrman, M.D.

CHAPTER ONE

The First Step—
Understanding Diabetes

*Jane Gillian was an obese fifty-six-year-old when she became
seriously ill and was hospitalized. She experienced an em-
bolic stroke, paralyzing her left side, and, while at the hospital, they also found that she had severe diabetes. Jane had a
family history of diabetes; both parents were overweight and
diabetic. Her medical history included high blood pressure,
high cholesterol, and placement of two medicated stents in
her coronary arteries. When she was admitted to the hospital with an HbA1C of 9.6 and a blood pressure of 200/100,
Jane was on two blood pressure medications as well as other
prescription pills. She was placed on insulin and remained in
the hospital for almost a month. Finally, she was discharged
wheelchair bound and on two insulin injections a day for a
total of 60 units daily plus eight other medications including
three blood pressure–lowering medications.*

A friend recommended Jane read Eat to Live, *and one
month later, she started the nutritarian diet. Her insulin
needs soon tapered and then stopped. Her results on the high-
nutrient diet were exciting. Three years later, Jane has lost
a total of 117 pounds—her weight went from 248 to 131*

pounds. Her HbA1C and glucose levels are in the nondiabetic
range. She is no longer diabetic. Her cholesterol dropped from
219 to 152, triglycerides from 174 to 66. Her blood pressure,
which used to run around 160/80 on the two blood pressure
medications, now runs around 125/75 without any blood
pressure medications. The best news of all is that Jane is no
longer in a wheelchair and can walk on a treadmill set at a
fifteen-degree incline for more than fifteen minutes.

Diabetes mellitus is a chronic disease that causes serious health complications including renal (kidney) failure, heart disease, stroke, and blindness. As mentioned, this serious disease has seen a drastic increase in the number of Americans who are affected. The Centers for Disease Control and Prevention released a 2011 report stating that over 25 million Americans are currently plagued by diabetes. That's an increase of 15 percent, or 3 million people, in only two years and over 700 percent in the last fifty years. More than 40 percent of Americans aged twenty years and older have either diabetes or pre-diabetes according to a review of data from the 2005–2006 National Health and Nutrition Examination Survey. Approximately 30 percent of adults older than sixty have been diagnosed with diabetes, and its prevalence is the same in men and women.

Many people are either unaware that they are diabetic or are in a pre-diabetic state that will lead to diabetes within a few years. The standard American diet (SAD) causes susceptible individuals to develop diabetes. Unfortunately, most people in America are eating themselves into a premature grave. The American diet is at the core of our health care crisis, and diabetics suffer even more tragic medical complications, such as:

- Heart disease—Death from heart disease and risk for stroke is three times higher for diabetics.
- High blood pressure—75 percent of diabetics have high blood pressure (130/180 or higher).

- Blindness—Diabetes is the leading cause of new cases of blindness among adults.

- Kidney disease—Diabetes is the leading cause of kidney failure.

- Nervous system disease—The majority of diabetics develop nervous system impairment such as reduced feeling in the feet, impaired digestion, and erectile dysfunction.

- Amputations—Diabetes is the leading reason for limb amputations.

- Cancer—Diabetes increases the risk of cancer, including a 30 percent increase in colorectal cancer.[1]

Diabetes is also taking a huge financial toll on America. Our unhealthy eating habits may eventually bankrupt our nation. The average type 2 diabetic incurs $6,649 in health care costs directly attributable to diabetes per year.[2] More than half of Americans will have diabetes or be prediabetic by 2020 at a cost of $3.35 trillion to the U.S. health care system if current trends go on unabated, according to analysis of a report released by UnitedHealth Group. Diabetes and prediabetes will account for the largest percent of health care spending by the end of the decade at an annual cost of almost $500 billion—up from an estimated $194 billion in 2010 according to the report titled *The United States of Diabetes: Challenges and Opportunities in the Decade Ahead.*[3]

In order to prevent this, we have to change the way we approach diabetes—and we must emphasize prevention. Earlier this year, the editors of the *Lancet* medical journal called it a "public health humiliation" that diabetes, a largely preventable disease, has reached such epidemic proportions. In reference to this year's ADA national meeting, the journal reported, ". . . there is a glaring absence: no research on lifestyle interventions to prevent or reverse diabetes. In this respect, medicine might be winning the battle of glucose control, but is losing the war against diabetes."[4]

These authors are correct—this is a public health humiliation because type 2 diabetes is both preventable and reversible. The SAD of refined grains, oils, sugars, and animal products is at the root of the crisis. Using drugs to keep glucose under control in individuals who continue to consume this diet will not prevent diabetes complications. The cure for type 2 diabetes is already known—removing the cause can reverse the disease.

Understanding the Cause

Every cell in the human body needs energy in order to function. The body's primary energy source is glucose, a simple sugar resulting from the digestion of foods containing carbohydrates (sugars and starches). Glucose from the digested food circulates in the blood as a needed energy source for our cells.

Insulin is a hormone produced by the beta cells in the pancreas, an organ located behind the stomach. Insulin bonds to a receptor site on the outside of cells and acts like a key to open a doorway into the cell through which glucose can enter.

When there is not enough insulin produced or when the doorway no longer recognizes the insulin key, glucose stays in the blood rather than entering the cells. So diabetes is the rise of glucose in the bloodstream due to a relative lack of the insulin that is responsible for the transfer of glucose from the blood into the tissues or cells. Normally as we eat and the glucose rises in the bloodstream, insulin-producing cells in the pancreas sense the glucose rise in the bloodstream. They then secrete the appropriate amount of insulin to drive the glucose into the body's tissues, lowering the level in the bloodstream back to an appropriate range.

Blood sugar	greater than 125	= diabetic
Blood sugar	110–125	= prediabetic
Blood sugar	95–110	= not ideal

When a person has type 2 diabetes, the amount of insulin produced is insufficient to lower the glucose level to normal; the level of glucose in the blood remains too high. In type 1, or juvenile, diabetes, the beta cells in the pancreas have been destroyed, so the body does not produce insulin at all. In type 2, or adult-onset diabetes, usually the body is not adequately responding to the insulin being produced. Fat on the body coats the cell membranes and impedes insulin function. The pancreas produces more and more insulin in response, but over time as the pancreas struggles with the extra workload, it eventually loses the fight and becomes unable to meet the unnaturally high demands. As insulin production starts to falter under the increased demands, the glucose in the bloodstream starts to rise. In both cases, with type 1 or type 2, insulin lack or insulin insensitivity, the glucose rises in the bloodstream. If it gets high enough, it also spills over into the urine. Initial symptoms of diabetes include frequent urination, lethargy, excessive thirst, and hunger.

The body will attempt to dilute the dangerously high level of glucose in the blood, a condition called hyperglycemia, by drawing water out of the cells and into the bloodstream in an effort to dilute the sugar and excrete it in the urine. It is not unusual for people with undiagnosed diabetes to be constantly thirsty, drink large quantities of water, and urinate frequently as the body tries to get rid of the extra glucose. This creates high levels of glucose in the urine.

Saving the Life of Type 1 Diabetics

Only about 10 percent of diabetics are type 1, also called childhood onset (or juvenile) diabetes because it typically begins in childhood. Type 1 diabetes refers to a disease in which the beta cells in the pancreas that produce insulin are destroyed by the immune system, usually early in life. When the body's immune system mistakenly targets our own cells instead of a foreign substance, it is called an autoimmune reaction. The causation is complicated and comes about partially as a result of an antibody reaction against a viral protein that mistakenly attacks the beta cells in the pancreas.

In this form, the body produces almost no insulin. It is characterized by a sudden onset and occurs more frequently in populations descended from northern European countries compared to those from southern European countries, the Middle East, or Asia. Type 1 is also called insulin-dependent diabetes because people who develop this type need to have daily injections of insulin.

Approximately 80 percent of our at-rest energy is used by the brain. Under normal situations, the brain can only function on glucose; however, when there is insufficient insulin, the brain and other tissues are unable to utilize the glucose in the bloodstream. When the body is unable to utilize glucose stores normally, free fatty acids will rise in the bloodstream. The body can make ketones from these fats, and then the brain and heart can use the ketones as an emergency fuel, when unable to get sufficient glucose. Glucose and ketones build up in the blood and can have devastating consequences. For example, type 1 diabetics are more prone to developing ketoacidosis, which can be life threatening if left untreated, leading to coma and death. Ketones are moderately elevated in blood and urine during fasting or significant carbohydrate restriction, but they can get to dangerously high levels in decompensated or untreated type 1 diabetes. Ketosis (high ketones in the blood) and ketoacidosis can occur in type 2 diabetics in some circumstances as well. It is the combination of the high glucose level in the blood along with the high level of ketones that can lead to dangerous acidosis and dehydration.

Type 1 diabetes is not caused by weight gain or obesity, and people with type 1 diabetes will always require insulin to prevent serious issues with high blood sugar (hyperglycemia) and other life-threatening conditions. Even so, a superior nutritional diet is essential for health and longevity of a type 1 diabetic, and even though excess body fat is dangerous for everyone, it is more dangerous for the type 1 diabetic.

I am often asked, "Is your program appropriate for type 1 diabetics? Will insulin be required forever, no matter what?" The answer to both questions is yes. Unlike a type 2 diabetic, if you are a type 1 diabetic, you can never stop taking insulin entirely. However, after adopting this high-nutrient dietary approach, you will need much less insulin, in most cases about half as much as before, following the typical ADA approach. The need for less insulin is not the only major reason for type 1 diabetics to follow this diet style. The vital reason is that it can save a type 1 from serious health complications later in life.

I have helped several patients with type 1 diabetes completely recover from their condition by flooding their body with micronutrients, fortifying their immune system, and resting the pancreas. This opportunity, however, is only available when the disease is just starting, usually in an adolescent or young adult. This is the exception, not the rule. Unfortunately, most type 1 diabetics have to live with the disease for the rest of their lives.

But here's the important news: With conventional care, the long-term outlook for a type 1 diabetic is dismal. More than one-third of all type 1 diabetics die before the age of fifty. This does *not* have to be the case. Type 1 diabetics need not feel doomed to a life of medical disasters and an early death sentence. Type 1 diabetics can lead a normal life and have a better-than-average life expectancy. It is true that type 1 diabetics are more sensitive to the damaging effects of the SAD diet, but if they eat a vegetable-based diet with plenty of beans, nuts, and seeds, they are no longer at risk for heart disease.

Scientific studies reveal that death due to early-onset heart disease in type 1 diabetics is linked to insulin resistance. That means weight gain, poor dietary choices, and therefore the need for excessive amounts of insulin is dangerous for type 1 diabetics. But when type 1 diabetics follow my nutritional advice, they require substantially less insulin and take it in physiologic dosages—the amount of insulin will not be excessive and will not hurt them.

Type 1 diabetics can have healthy, normal, and long lives. The typical health tragedies that befall type 1 diabetics are the result of the combustible combination of American food and excessive insulin use, a fire fueled by physicians and dieticians whose nutritional advice unfortunately remains in the dark ages.

By adopting this high-nutrient approach, type 1 diabetics lower their insulin needs and no longer have swings of highs and lows. Glucose levels and lipids stay under control with minimal insulin. Requiring less insulin while still having excellent glucose readings is the goal. The simple truth is that the reason why type 1 diabetes leads to heart attacks and other life-shortening ailments is the excess insulin required by a low-nutrient diet, not the diabetes itself.

It is not type 1 diabetes that causes such negative health consequences. Rather, it is the *combination* of the diabetes and the typical nutritional "advice" given to patients—advice that requires them to take large nonphysiological amounts of insulin to maintain favorable glucose readings. Insulin itself promotes the development of atherosclerotic plaque, the foundation of heart disease and heart attacks. Insulin increases appetite and promotes fat storage and weight gain, thus furthering insulin resistance. This is particularly exacerbated by the high glycemic and excessive caloric load in conventional diets.

I have been on your plan for two years and am really happy with the results. I am at my ideal weight with about 10 percent body fat. A couple of years ago I was 190 pounds with high cholesterol. My insulin was at 30u Lantus and Humalog on a sliding scale but often like 6u per meal. Following your advice I dropped the weight to 170, my cholesterol is awesome now, and blood pressure and lipid profiles are great! Now my Lantus is 10u and I am on Novolog, two or three units per meal.

When I was diagnosed in my teens, my doctor said there were two ways to look at the diagnosis:

1. as the end of my health forever or

2. an opportunity to gain an understanding of my body and how it works and become healthier than ever

I tried to take the latter road, and now, at age thirty-four, I think I am finally realizing that potential. Your writings were the suit of armor I needed in the fight all these years. Thanks again for everything.

—Tony Gerardo

Several studies illustrate the dangers of giving insulin to the adult diabetic. In one such study, when diabetic patients were given insulin, compared to those given metformin (Glucophage), the risk of death from heart attacks tripled.[5] The negative effects of insulin are related to both the systemic metabolic abnormalities from excessive insulin and the direct pro-atherogenic effects of insulin on the endothelial lining of blood vessels that promotes atherosclerosis.[6] The more insulin that is needed, the more dangerous plaque is promoted, especially when the amount of circulating insulin is high. Extra insulin and high blood sugar levels also raise cholesterol, promote fat deposition, and damage the body. With this in mind, it should be clear that while the SAD, which has spread to all industrialized nations, is dangerous for everyone, it is particularly deadly for diabetics. Diabetes is not a death sentence, but we can't keep following conventional medicine and dieticians' advice or the excessive insulin and overuse of other medications they call for.

The negatives of overprescribing insulin are not limited to weight gain and heart disease. The connection between diabetes and cancer is thought to be due at least in part to insulin therapy. A new review that analyzed data from several studies found that diabetic patients are 30 percent more likely to develop colorectal cancer, 20 percent more likely to develop breast cancer, and 82 percent more likely to develop pancreatic cancer.[7] I am certain that by using insulin in small physiological amounts in type 1 diabetics, whose insulin needs would be low on my nutritarian diet, the metabolic negatives and the increased risk of cancer from insulin would *not* be noted. These negatives are the result of the excessive use of insulin necessitated by the SAD and the standard diabetic diet.

When type 1 diabetics follow the Eat to Live approach, it is possible to prevent many of the complications that can accompany the disease. As discussed, a normal life and life span are well within reach. Type 1 diabetics will still require insulin, but for almost all patients, the insulin dosages required will be greatly reduced, and they will require only the amount of insulin that a person's pancreas would secrete if eating healthfully and nondiabetic, so no damage will ensue because they are not requiring abnormally high amounts of insulin.

Specifically, if type 1 diabetes is well managed, there will be many benefits:

- No highs or lows in blood sugar
- Less insulin use—most typically, dose is cut by half
- Normal, stable body weight
- Normal life span, without diabetic complications

The key formula to remember here is that favorable glucose levels + excellent nutrition = a healthy and long life. If you or someone you love has type 1 diabetes, please read this book. I promise that it can save lives; I have seen it happen.

The Dramatic Increase in Type 2 Diabetes: A Tragic Phenomenon

Type 2 diabetes occurs in approximately 3 to 5 percent of Americans under fifty years of age and increases to 10 to 15 percent in people over fifty. More than 90 percent of diabetics in the United States are type 2 diabetics. Sometimes called adult-onset diabetes, this form of diabetes occurs most often in people who are overweight and who do not exercise sufficiently. The explosion in the occurrence of diabetes in the last twenty-five years in America parallels the skyrocketing number of overweight people.

Type 2 diabetes almost never occurs in people who eat healthy, exercise regularly, and have a low body fat percent. The disease hardly existed in prior centuries when food was not so abundant or when high-calorie, low-nutrient food was not available. It is also more common in people of Native American, Hispanic, Indian, and African-American descent, though no background is immune to the effects of a diabetes-inducing diet. Worldwide, diabetes is exploding as populations in all corners of the globe are being exposed to processed foods for the first time in human history. The development and abundance of processed foods in the world's food supply combined with more sedentary jobs has created an explosion of obesity, diabetes, and heart disease. Most countries have attempted to solve this problem with medications for diabetes, high blood pressure, and high cholesterol. Invasive medical procedures and surgeries are used at a substantial expense but without significant life span enhancements or benefits to society.

In the United States, being overweight is the norm, and almost all adults eventually take prescribed medications for their heart, diabetes, cholesterol, or blood pressure. In fact, 51 percent of those over the age of 65 take five or more prescription drugs a day! The number of obese Americans is higher than the number of those who smoke, use illegal drugs, or suffer from other physical ailments. Obesity is a major risk factor associated with highly prevalent serious diseases such as heart disease, cancer, and diabetes. It is what we eat that creates these diseases and fuels out-of-control medical costs. Even five extra pounds on a normal body frame can lead to diabetes.

Research shows that excess body fat is the most significant cause of type 2 diabetes. Through working with thousands of patients, I have observed with consistency that losing body fat in conjunction with maintaining high levels of micronutrients in the body's tissues will reduce the need for medications and, in most cases, reverse type 2 diabetes for good. As we'll explore in detail throughout this book, scientific studies show it is not just the weight loss but also the cell's exposure to a favorable micronutrient environment that enable recovery. Many

Obesity Trends Among U.S. Adults

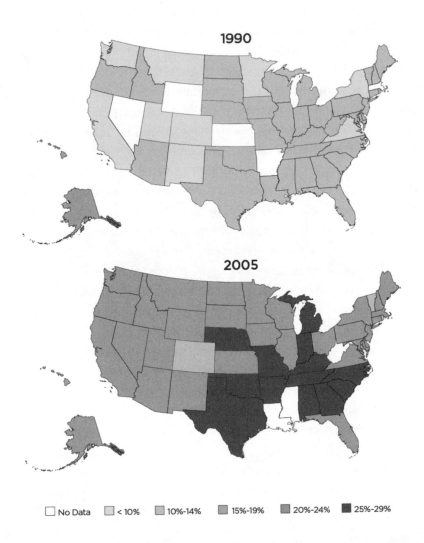

1990

2005

| ☐ No Data | ☐ < 10% | ☐ 10%-14% | ☐ 15%-19% | ☐ 20%-24% | ■ 25%-29% |

of my patients recover from their diabetes before most of their weight has been lost. The cells become more responsive to insulin when the body is not burdened with excess fat, and the high level of micronutrients in the tissues enables the beta cells that have pooped out from struggling to produce extra high levels of insulin for years to reclaim lost function.

Because of its slow onset and the fact that it can usually be controlled with diet, type 2 is considered a milder form of diabetes, sometimes developing over the course of several years. The consequences of uncontrolled and untreated type 2 diabetes, however, are just as serious as those for type 1. Heart attacks, infections, amputations, blindness, and strokes are possible, but unlike type 1, type 2 diabetics can almost all come off insulin and other medications if they take off the excess weight.

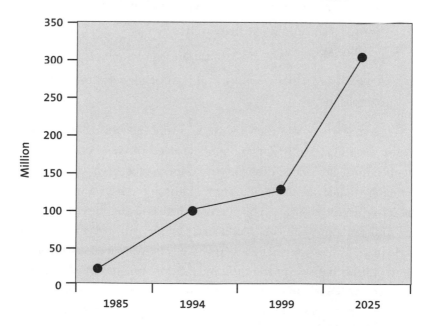

Prevalence of Diabetes Worldwide

Diabetes isn't just about elevated blood sugar levels—which pose immediate threats including blurred vision, drowsiness, confusion, and vomiting—it's about every other long-term condition and complication it creates as well. It can take a severe toll on the health of a diabetic—increasing not only the risk of heart attacks and strokes but also of depression and cancer.[8]

What a Type 2 Diabetic Can Expect

Specifically, if type 2 diabetes is well managed with exercise and superior nutrition, there will be many benefits:

- No highs or lows in blood sugar
- Reduction of medications by an average of 50 percent in the first week, more in the first month, and most typically 100 percent within six months
- Need for insulin is eliminated, usually within the first week
- Normal, lean, and stable body weight
- Normal life span, without complications
- Reversal of diabetes and prevention of diabetes-related complications

The goal is to reverse diabetes to the point of becoming nondiabetic again, meaning ideally that your glucose levels run below 100 without medications. Be aware, though, that once you've been diabetic, the tendency to become diabetic again remains if you regain weight or go back to unhealthy eating. This is a new diet style and lifestyle forever.

You can anticipate your blood sugar falling with this diet and lifestyle plan. As discussed earlier, you will be able to reduce your medications. Err on the side of too little medication, not too much. Prevent the occurrence of hypoglycemic episodes with good communication with your physician and careful use of minimal medications.

If your blood glucose has been elevated for a while, even as your blood sugar approaches the normal range, you could feel somewhat ill as the body gets accustomed to experiencing normal blood glucose levels. Nevertheless, when on diabetic (glucose-lowering) medications, especially insulin and sulfonylureas—Amaryl (glimipiride), Diabenese (chlorpropamide), Glucotrol (glipizide), Diabeta, Glynase (glyburide), Actos (pioglitazone), Avandia (rosiglitazone)—it's important to check your blood sugar frequently during the first week to make sure you are not being overmedicated. Glucophage (metformin) is a commonly used oral diabetes medication that does not cause the blood glucose to drop too low and does not cause weight gain, so this is the preferred medication to remain on, if one is needed.

Snacking to prevent a hypoglycemic reaction from the overuse of medication is poor medical management and should not happen. Medications should be reduced in time so this never occurs. I tell patients starting this program that if a blood sugar reading is below 120, it is time for the next round of medication reduction. It is better to be undermedicated slightly, to prevent the need to treat hypoglycemic events, than it is to be overmedicated. If the diabetic patient experiences hypoglycemic episodes and extra snacking is required to bring the glucose up, then the physician overmedicated the patient and did not do his job correctly.

The ADA diet uses the diabetic exchange list to help diabetics create what they call balanced meals. This exchange diet divides foods up into groups based on similarities in nutrient content and includes starches, fruits, milk, vegetables, meats, fats, sweets, and other carbohydrates. It looks to make meals that are based on a preconceived notion that balancing an equal amount of fat, carbohydrates, and protein at each meal is favorable. It then allows exchanges based on the amount of calories from that macronutrient. For example, in the starch group, one slice of toast can be exchanged for a half cup serving of cooked oatmeal.

Because the foods the diet is designed with are inherently poor in fiber, micronutrients, and resistant starch, they fuel an obsession with food because the dieter is never satisfied. This continual struggle with dieting and trying to maintain small portion sizes of foods that do not biologically fill you up rarely works. Even in controlled dietary studies in which calories are carefully monitored, the results are relatively poor simply because the American dietary standard is so poor and the ADA diet mimics this failed dietary pattern utilizing too much unhealthy, low-micronutrient food. Researchers have also frequently noted the difficulties involved in the ADA plan, particularly the requirements to dramatically restrict portion sizes that most individuals simply cannot comply with long term.[9]

An ADA sample breakfast meal may include two slices of toast with one teaspoon of margarine, a scrambled egg, three-quarters of a cup of unsweetened ready-to-eat cereal with one cup of nonfat milk, and a small banana. Another breakfast choice on the 1,800-calorie ADA diet may include two four-inch whole wheat pancakes with two table-spoons of light pancake syrup, one teaspoon of margarine, one cup of sliced strawberries, one-quarter cup of low-fat cottage cheese, and one cup of nonfat milk. These sample meals are a formula for disaster for diabetics. In order to get the glucose controlled after consuming all those low-fiber carbohydrates, an excessive amount of diabetes medi-cation will have to be prescribed, which will lead not only to highs and lows but also potentially to hypoglycemic episodes. Then diabetic patients are instructed to snack to prevent the low blood sugar results of the medication, further impeding their possibility of dropping the excess body fat. The additional side effects and weight gain from the medications just lead to a worsening of the diabetes. The focus with standard care is on the glucose level and maintaining the right amount of medication to optimally stabilize the glucose. It misses the boat, though, because it fails to focus on the health and weight of the person first, and the miraculous health and weight loss benefits of the right dietary pattern based on greens, beans, mushrooms, onions, tomatoes, peppers, berries, intact grains (not just whole grains), seeds, and nuts.

In contrast, type 2 diabetics can become nondiabetic, achieving complete wellness and even excellent health. They can be diabetes-free for life. In my twenty years of clinical experience with this program, I have experienced that more than 90 percent of type 2 diabetics who follow this diet and exercise lifestyle are able to discontinue insulin within the first month.

CHAPTER TWO

Don't Medicate, Eradicate

Jim Kenney, a fifty-eight-year-old male, was referred to my office from his nephrologist (kidney specialist) at St. Barnabas Hospital in Livingston, New Jersey. He was originally referred to the nephrologist by his endocrinologist (diabetic specialist) at the Joslin Diabetes Center in Boston because of kidney damage that resulted from very high glucose readings in spite of maximum medical management. At this first visit, Jim weighed 268 pounds and was on 175 units of insulin per day (a very high dosage). He had already suffered from severe complications of type 2 diabetes, including two heart attacks and Charcot (destructive inflammation) joint damage in his right ankle. In spite of this huge dose of insulin and six other medications, Jim's glucose readings averaged between 350 and 400. Jim said this was the case no matter what he ate, adding that he was already on a diabetes diet and was already following the precise diabetes nutrition and dietary recommendations of the dietician at the Joslin clinic.

During his first visit with me, after we discussed his new diet program, I reduced his insulin dose to 130 units per day. The following few days, Jim and I spoke over the phone, and I continued to decrease his insulin gradually. Within five days, Jim's glucose was running between 80 and 120 and he lost ten pounds. At this juncture I reduced his Lantus long-acting insulin dose to 45 units at bedtime and his Humalog regular pre-meal insulin to 6 units per meal, for a total of 63 units per day.

At his two-week visit, Jim had lost sixteen pounds. I was already stopping some of his blood pressure medications and he was down to a total of 58 units of insulin per day. After the first month, I was able to stop all of Jim's insulin and start him on Glucophage (metformin). He lost twenty-five pounds in the first five weeks, and his blood glucose readings were well controlled without insulin. In addition, his blood pressure came down to normal, he no longer required any blood pressure medications, and his abnormal kidney function was improving. Five months later, Jim had lost sixty pounds and was off all medications for diabetes. He no longer had high cholesterol or high blood pressure. His kidney insufficiency had completely normalized as well.

Jim's story illustrates not merely how powerful this dietary protocol is but also how the standard nutritional advice given to diabetics from conventional physicians and dieticians can be disease promoting. The standard nutritional advice given to diabetics is not only insufficient—it is dangerous. Jim Kenney would likely be dead by now had his nephrologist not referred him to my office.

To begin examining how type 2 diabetes can be healed, we need to look at how it developed in the first place. As mentioned, the heavier you are, the greater your risk of developing type 2 diabetes. For some people even a small amount of excess fat on the body can trigger diabetes. Your body's cells are fueled primarily by glucose. Insulin is the hormonal messenger produced by the beta cells in the pancreas, which induces glucose uptake into the body's cells. Glucose cannot pass into the cells unless insulin opens the gate. However, as little as five pounds of excess fat on your frame can interfere with insulin's ability to carry glucose into your cells. When you have excess fat on your body, insulin does not work as well, and then the glucose has difficulty entering the cells. Fat on the body interferes with the action of insulin through multiple mechanisms.

Free fatty acids released from the fat cells is one of the mechanisms promoting insulin resistance in liver and muscle in a phenomenon known as lipotoxicity. The excess of circulating fats in the bloodstream also blocks insulin binding on the outer membrane of cells and interferes with normal muscle cell function and energy production. When cellular energy production is slowed, more insulin is required. This lipotoxicity can affect the heart as well, promoting an irregular heart beat and increasing susceptibility to heart failure.

Fat cells also produce binding proteins that attach to the insulin hormone blocking its activity. Some of these fat cell–produced molecules also cause muscle cells to be desensitized to insulin. If that is not bad enough, when our cell membranes are impregnated with dietary trans fats and saturated fats, the insulin-binding sites are distorted, impairing insulin from binding to the docking station on the cell membranes, making insulin less effective at enabling glucose uptake. To overcome all these issues, your pancreas must produce additional insulin. With significant weight gain, the insulin-producing beta cells in the pancreas become dramatically overworked. In short, type 2 diabetes is a disease of heightened insulin *resistance,* not one of absolute insulin *deficiency.*

Insulin works less effectively when people eat fatty foods, overeat, eat low-nutrient foods, or gain weight. So when people are overweight, they require more insulin, whether they're diabetic or not. But giving overweight diabetic people even more insulin makes them sicker by promoting further weight gain, causing them to become even more diabetic. How does this process work? Our pancreas secretes the amount

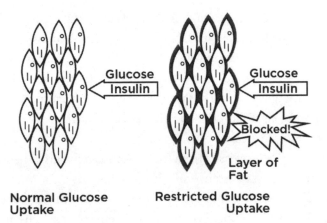

Normal Glucose Uptake

Restricted Glucose Uptake

of insulin demanded by the body. People of normal weight with about one-third of an inch of periumbilical fat will secrete a certain amount of insulin. But what happens when they gain twenty pounds of fat? Their bodies will now require more insulin, almost twice as much, because the fat on their bodies interferes with the uptake of insulin into the cells by the various mechanisms mentioned.

Body Fat Deactivates Insulin and Raises Blood Glucose

- Free fatty acids circulating in the blood have a toxic effect, inhibiting energy production from muscle tissues, which then demand more insulin.

- Fat cells produce pigment epithelium-derived factor, causing cells to be desensitized to insulin.[1]

- Fat cells produce retinol-binding protein, which prevents insulin from activating glucose-carrying proteins.[2]

- Trans fats and saturated fats can stiffen and distort membrane-located insulin receptors, interfering with efficient binding.[3]

When people are significantly overweight or obese, with more than fifty pounds of excess fat weight, their bodies demand huge loads of insulin from the pancreas, even as much as ten times more than people of normal weight require. What do you think occurs after ten or more years of overworking the pancreas so hard? Of course, it becomes exhausted and loses the ability to keep up with the huge insulin demands, and less insulin is produced. Eventually, with less insulin available to move glucose from the bloodstream into the cells, the glucose level in the blood starts to rise, and those people are diagnosed with diabetes. In most cases, these people are still secreting an excessive amount of insulin, compared to normal-weight people, but just not enough for them. As time goes on, even though the overworked pancreas may still pump out much more insulin than a thinner person might need, it won't be enough to overcome the effects of the disease-causing body fat. I call it pancreatic poop out.

Some severely overweight individuals have a large pancreas beta cell capacity, so they can produce high levels of insulin without becoming diabetic. These high insulin levels in the blood are a strong predictor of heart attack risk and shortened life span. So whether these people are diabetic or not, their high insulin levels are still dangerous. In fact, insulin level is a better indicator of a future heart attack than cholesterol level. Often people will be in an emergency room with their first heart attack and be told for the first time that their sugar is elevated. These heart attack victims never knew they had diabetes. The first sign of it was the heart attack from years of having a heightened insulin level. Damage was building up before the elevated glucose became apparent.

In most cases, the pancreas's ability to produce insulin continues to lessen as the diabetes and the overweight condition continue year after year. Unlike type 1 diabetes, total destruction of insulin-secreting ability almost never occurs in type 2 diabetes. But the sooner type 2 diabetics lose the extra weight that is causing the diabetes, the greater the likelihood they will be able to maintain a functional reserve of insulin-secreting cells in their pancreas.

What this means is that typical type 2 diabetes is caused by excess weight in individuals who have a smaller reserve of insulin-secreting cells in the pancreas. As the statistics are showing, type 2 diabetes is a growing epidemic. But what is surprising is that people suffering can range anywhere from ten pounds overweight to significantly obese. It is important to say here that in individuals who are susceptible, ten to twenty extra pounds can lead to diabetic symptoms. No matter what the number is, losing the excess weight enables these individuals to live within the capabilities of their body. Most type 2 diabetics still produce enough insulin to maintain normalcy as long as they maintain a favorable body-fat percentage.

Simply put, since the level of insulin in your blood is a good indicator of your risk for heart attack, and since a tape measure around your waist is nearly as good an indicator of insulin levels as a blood test, it makes sense to remember the ancient saying, "The longer your waistline, the shorter your lifeline."

Following a nutrient-rich, lower-calorie diet—a nutritarian diet—coupled with a good exercise program is the most important change you can adopt to extend your life span. It has been known for years that intentional weight loss improves blood sugar, lipids, and blood pressure in diabetics. Gastric bypass surgery and lapband procedures are risky, lead to malnutrition, and most often produce only temporary results. Nevertheless, overweight individuals who go through gastric bypass surgery and become too uncomfortable to eat much often also resolve their diabetes. Over the years, as the stomach stretches and the weight returns, these individuals can become diabetic again. Unfortunately, they did not learn enough about nutritional excellence. A recent study documented a significant increase in life span, with an average of 25 percent reduced mortality, when diabetic individuals dropped their body weight by just twenty to twenty-five pounds.[4] Imagine the results when a program of nutritional excellence achieves the weight loss and the body's cells are flooded with micronutrients that fuel cellular repair. Scientific literature shows it is not just the weight reduction that enables diabetic reversal and recovery but also the high level of plant-derived micronutrients and phytochemicals that can fuel the body's own remarkable self-healing properties.[5]

The results you can achieve with a nutritarian diet are predictable and remarkable, but it takes some effort and time. There are lots of diet books and exercise plans written for diabetics, but this nutritarian diet is designed and proven in clinical practice to be the most effective for losing weight, lowering cholesterol, and reversing diabetes. It is the gold standard, written specifically for people who want to do what is very best for their health and give it their all to become nondiabetic.

A nutritarian approach is all about superior nutrition, not just moderate improvement in diet. Moderation doesn't work. But not to worry—as already mentioned, nutritional excellence will make your taste buds happy and you will be more than satisfied with the amount of food you can eat. But we will get into that later in the book.

Decreasing Insulin and Other Medications

Type 2 diabetics are overweight to begin with and, as you have learned, being overweight is the significant causative factor in diabetes. Because insulin therapy results in further weight gain, how could giving more insulin or oral medication to force the already overworked pancreas to produce more insulin be a good thing? A vicious cycle begins that usually causes diabetics to require more and more insulin or other medications as they put on the pounds. On their initial visit to me, patients often report their sugars are impossible to control in spite of massive doses of insulin, which are typically combined with oral medication. These patients are significantly overmedicated but are still overweight and eating unhealthfully. It is like they are walking around with a live hand grenade, ready to explode at any minute.

Excess insulin in the same environment as excess weight, high cholesterol, hypertension, and inflammation from inferior micronutrient exposure promotes hardening of the arteries, which will eventually lead to heart attacks and strokes. Studies have shown that high levels of insulin in the blood promote hardening of the arteries even in nondiabetics. In diabetics, the effects of excess insulin are even worse. In a study of 154 treated diabetics, blood vessel disease was greatest in those with the highest levels of insulin.[6] It made no difference whether the insulin was self-produced in the body or taken by injection. Quite a few studies illustrate the dangers of giving insulin to type 2 diabetics. When these patients are given insulin—compared with those given an oral antidiabetes medication, the risk of death from heart attacks tripled.[7]

The bottom line is that insulin use creates a vicious cycle that cuts years off a person's life. Insulin both blocks cholesterol removal and delivers cholesterol to cells in the blood vessel walls, increasing the risk for heart attacks and strokes. Almost 80 percent of all deaths among diabetics are due to hardening of the arteries, particularly coronary artery disease. Many diabetics turn to their physician for guidance, but oftentimes the well-meaning doctor only worsens the problem by prescribing more insulin. The extra insulin does not just cause heart disease, weight gain, and the eventual worsening of the diabetes; as with type 1 diabetes, insulin can increase the risk of cancer as well. Type 2 diabetic patients exposed to insulin or sulfonylureas, which push the pancreas to produce more insulin, have significantly increased incidence of cancer at multiple sites.[8]

Many other unfavorable side effects occur from using diabetes medications. For example, medications such as insulin and thiazolidinediones like Actos and Avandia not only cause weight gain and leg swelling but also, as reported in the April 2009 issue of the *American Journal of Ophthalmology*, have been shown to dramatically increase a diabetic's risk of developing macular edema, a serious eye disease. Recently, a study published in the *British Medical Journal* examining over ninety thousand diabetics demonstrated a significantly higher risk of heart failure and all-cause mortality (death) in diabetic patients prescribed sulfonylureas.[9] Sulfonylureas are one of the most commonly prescribed drugs for diabetes. A recent retrospective study, reported at the 2012 annual meeting of the Endocrine Society, reviewed these widely prescribed diabetic drugs in 23,915 patients with type 2 diabetes on monotherapy (one medication only). It reported the death rates on patients taking glipizide, glyburide, or glimepirmide (all sulfonylurea drugs) and found they had a 58 to 68 percent increase in all-cause mortality compared to patients taking only metformin. This study may have under-represented the dangers since it only followed the patients for 2.2 years.[10]

Clearly our present dependency on drugs to control diabetes without an emphasis on dietary and exercise interventions is promoting diabetic complications and premature death in millions of people all over the world.

The tendency to throw drugs at every medical condition is the problem with medicine today. Physicians prescribe drugs in an attempt to lower dangerously high blood sugar, risky high cholesterol, and damaging high blood pressure levels typically seen in diabetics, since these high levels can lead to further damage or premature death. Unfortunately, treating diabetes with medication gives patients a false sense of security because they mistakenly think their somewhat controlled glucose levels mean they are healthy. Whether patients have high cholesterol, high blood pressure, or any other risk factor, the use of medication takes the emphasis away from the complete overhaul of the lifestyle and diet style that is absolutely essential to save their life. Going to doctors and getting a pill for every issue has a subconscious effect to avert personal responsibility, and the motivation for patients to earn back their health is lessened. This provides diabetics (and heart patients) with the justification to continue with the same disease-causing diet and lifestyle that led to the development of their condition in the first place, while falsely believing they are receiving significant protection.

What patients (and many physicians) do not understand is their "controlled" diabetes continues to damage their organs and heart. Inevitably, the diabetes worsens, tragic complications develop, and patients die much too soon. Eighty percent of adults with diabetes die of heart attacks and strokes. Tragically, much of this suffering is unnecessary because diabetes and its complications can be avoided.

What's worse is that physicians often advise diabetics to learn to live with and manage their diabetes because they say it cannot be healed or cured. Type 2 diabetics who adopt a healthy nutritional approach *can defeat diabetes* and achieve excellent health. That's diabetes-free for life. Almost all of my type 2 diabetic patients are weaned off insulin within the first few weeks, and thanks to excellent nutritional habits, they have much lower blood sugar than when they were on insulin. Stopping insulin also makes it easier to lose weight.

What's a Doctor to Do?

Conventional physicians specializing in diabetes are in a bind. They know that high blood sugar levels create problems—not just by stressing the heart but also by aging the eyes and kidneys, leading to devastating complications such as kidney failure and blindness. They want to prescribe aggressive insulin therapy to decrease patients' blood sugar. The problem is, they also are aware that the extra insulin accelerates hardening of the arteries (which leads to heart attacks) and weight gain (which eventually makes patients *more* diabetic). Tightening blood sugar control with insulin is risky business. In fact, studies that follow patients who carefully monitor their glucose level, adjusting their medications precisely to maintain the most favorable levels, show that these people have increased mortality. They do not do better. The only way to beat diabetes is to get thin, eat right, and use less medication. The increased use of medications is to blame when diabetics attempt to maintain lower glucose readings and then die younger.

On February 6, 2008, the National Heart, Lung, and Blood Institute stopped the Action to Control Cardiovascular Risk in Diabetes study when results showed that intensive treatment of diabetics increases the risk of dying compared to patients who are treated less aggressively. When you read the comments of physicians and researchers discussing these results, it is apparent that they still do not understand why this occurred. Physicians are still looking for the magic combination of drugs to treat diabetes. They still do not understand that drugs cannot effectively treat this disease, which is merely a side effect of an unhealthy lifestyle and diet. Giving stronger and stronger drugs—which drive up appetite, cause more weight gain, and rack up other detrimental side effects—will never be the right approach for type 2 diabetes. No medications can do what a dietary and lifestyle overhaul can.

Most physicians would likely agree that weight reduction and high-nutrient eating is the most successful route to health, but they do not know much about it or how to motivate their patients to change, and they doubt their patients would do it. Certainly, in rare instances when physician interventions are successful at achieving significant weight reductions, the outcomes are invariably positive. We have already discussed that patients with diabetes who undergo gastric bypass surgery typically see their diabetes resolve.[11] Plus nutrition interventions that control and limit calories have been effective for reversing diabetes too, enabling many patients to discontinue medications.[12]

Preventing and reversing diabetes is not all about weight loss. The nutritional features of this diet have profound effects on improving pancreatic function and lowering insulin resistance over and above what could be accomplished with weight loss alone. The increased fiber, micronutrients, and stool bulk, plus the cholesterol-lowering and anti-inflammatory effects of this high-micronutrient eating style, have radical effects on type 2 diabetes. Scores of my patients have been able to restore their glucose levels to the normal range without any further need for medications. They have become nondiabetic. Plus, one's blood pressure, cholesterol, and overall health and vitality are radically improved or normalized. Even my thin, type 1, insulin-dependent diabetic patients are able to reduce their insulin requirements by about half. They experience greatly improved glucose control and stabilized highs and lows, which protects them from the typical dangers that are almost inevitable to long-term diabetics who eat more conventionally.

Sadly, the ADA as well as most dieticians and physicians offer dangerous advice to diabetics. They provide minimal guidance on weight reduction and cholesterol lowering, and worse yet, the diets they recommend are not successful for helping diabetics lose weight and keep it off. Typical diabetes care is focused on the wrong thing—monitoring blood sugar to determine when it is necessary to change insulin dosages or adjust other medications.

Instead of motivating excellent nutrition to prevent disease, the ADA reinforces our disease-causing food habits. For example, here are some statements from the ADA website:

Fact: If eaten as part of a healthy meal plan, or combined with exercise, sweets and desserts can be eaten by people with diabetes. They are no more "off limits" to people with diabetes than they are to people without diabetes.

Fact: For most people, type 2 diabetes is a progressive disease. When first diagnosed, many people with type 2 diabetes can keep their blood glucose at a healthy level with oral medications. But over time, the body gradually produces less and less of its own insulin, and eventually oral medications may not be enough to keep blood glucose levels normal. Using insulin to get blood glucose levels to a healthy level is a good thing, not a bad one.

This advice is flat-out wrong. Case in point with the latter fact: as diabetics are given inadequate dietary advice, placed on medications that cause weight gain and push the failing pancreas to work harder, and generally guided to mismanage their diabetes, the result will of course be more medication and the eventual need for insulin. This is simply drug-promoting double-talk that makes medications the answer over effective and proven lifestyle interventions. The ADA medical advisory committee states: "It is nearly impossible to take very obese people and get them to lose significant weight. So rather than specifying an amount of weight loss, we are targeting metabolic control." This is doublespeak for "Our recommended diets don't work, so we just give medications and watch patients deteriorate."

Physicians engaging in such conventional medical practice are endangering their patients' lives. Instead, they should always offer the option of treating diabetes with effective nutritional and dietary changes. The problem is that most physicians don't really understand the proper nutritional recommendations to make.

How can diabetics safely lower the high blood sugar levels that are slowly destroying their bodies? How can they lower their cholesterol and blood pressure, lose weight, and avoid taking dangerous drugs? The most effective glucose-lowering drugs are also the most dangerous in the long term.

The best medicine for diabetics is a high-nutrient, lower-calorie diet and exercise, not drugs. This is the only approach that lowers cholesterol, lowers triglycerides, and lowers blood pressure as it drops weight and blood glucose. High-nutrient plant foods also have an anti-inflammatory effect on blood vessels and organs. This enables self-repair mechanisms that are essentially disabled on a low-nutrient diet. This dietary approach has helped thousands of diabetics reduce or eliminate the need for insulin and other medications. It has changed the entire course of their health and longevity through the foods they eat.

The bottom line is this: you can get rid of your
diabetes, not just "manage it."

Do not rely on standard drug methods to treat diabetes. With no
medication to cover up their dietary failings, diabetics will be compelled
to eat properly and exercise more to control their elevated glucose. This
aggressive approach based on nutrient-rich foods is the most effective
way to reverse this dangerous condition. Learning about the nutrients
inside these healing foods is an important step in defeating diabetes.

Your Health Future (H) = Nutrients (N) / Calories (C)

A career in medicine is so much more rewarding when patients actually get well. How often does the physician say to his patient, "Congratulations, you do not need drugs for your high blood pressure and
cholesterol anymore. You did it! You removed your risk factors because
you are healthier!" Or, "Congratulations, your stress test has normalized." Or, "Your carotid ultrasound shows no visible plaque anymore."
These are typical statements I make every day in my office. It is exciting
to see people recover their health.

Unfortunately, too many people in this profession do not give their
patients the opportunity to get healthy. Imagine if all doctors told their
patients that diet and exercise are more powerful than drugs and they
were adamant about compliance. Instead, most physicians have the as-
sumption that the effort is too great; that patients are not willing eat right,
exercise, and get slim; and that drugs are the only answer. They prescribe
drugs and tout them as the only viable option and then watch patients'
health deteriorate. Often they don't know there is another, more effective
option. In all of this, the public loses. So much of our conventional medical system is based on ignorance. Nutritional medicine, when practiced
properly, is much safer and more effective than conventional medicine.

A medical diagnosis such as diabetes is an opportunity for physicians to teach patients what they are doing to hurt themselves and how

the American diet is disease causing. It is an opportunity to motivate them to earn back superior health. On every visit, the nutritionally astute physician should review the patient's list of medications and gradually be able to reduce dosages or discontinue medications.

Medications are an insufficient and ineffective intervention for the chronic diseases that have been created by bad lifestyle and dietary choices. A considerable part of the problem is bad information. Inappropriate diets of all descriptions flood the marketplace, and traditional dietary teachings are riddled with myths and inaccuracies. The failure of conventional dietary programs to achieve long-term weight reduction merely reflects the weakness of the advice given and the poor educational and motivational techniques offered. This lack of awareness—even among health professionals—does not weaken the science, logic, or effectiveness of utilizing superior nutrition and lifestyle interventions as the primary therapeutic modality, however.

Following a correct diet and exercise plan as a remedy should not be labeled alternative or complementary medicine. It is simply the way all properly educated doctors should be practicing. Everything else should be called malpractice medicine. Offering patients drugs and surgical interventions without informing them that, for most diseases, nutritional excellence and exercise are safer and more effective in the long run is not adequate informed consent to the use of medications. The risks of medicines are downplayed and their supposed benefits greatly exaggerated by a medical profession and drug industry who offer drugs as the panacea to all that ails us.

Most often alternative, or complementary, medicine offers the same treatment mentality as the physicians focused on dispensing drugs. Rather than dealing with the dietary and lifestyle factors that caused the condition, alternative physicians are also offering some magic in the form of an herbal pill or IV vitamin drip. Natural herbs or other modalities can sometimes offer similar effects as drugs. There are plenty of natural substances that have therapeutic effects, but they do not deal with the cause of the problem, so their benefits are limited. A diet-induced disease needs a dietary solution, not more treatment options. It is typical to find an alternative physician offering chelation, IV nutrients, hormones, and an expensive assortment of supplements and remedies while the patient remains fifty pounds overweight. Effective weight reduction will not be achieved. Remedies can change the expression of symptoms, but they never make patients well. They just cause people to become more dependent on doctors and their remedies. Physicians and consumers are quick to embrace doctor-recommended medical interventions while they ignore simple, inexpensive, and dramatically effective lifestyle interventions. Optimal lifestyle medicine would free these people from needing medical care. They need less and less therapy and medical intervention. This path of advice is not a moneymaker for the professional. There is no huge economic incentive to promote the basics of good health.

The problem with lifestyle medicine today is the varying opinions and dietary programs that are popular but not ideal. These in turn result in a few studies showing limited effectiveness—all of which help the current medical approach make its case for prescriptions.

Wouldn't it be simpler if we all could agree on one program? If there was one program that was most logical, most effective therapeutically, and beat out all the others when subjected to scientific scrutiny and long-term evaluation, it would change the health conversation radically, and all doctors and healers would naturally begin embracing such a protocol. I have been developing, teaching, and fine-tuning this program over the last twenty years. Studies continue to show it can meet any scrutiny and testing. It is ready for implementation and documentation. It works effectively for a surprisingly wide array of chronic diseases and does not have to be overly complicated. As we've explored, significant research already supports its use, and further research is presently in process. Test it for yourself, and you will be shocked with its effectiveness.

CHAPTER THREE

Standard American Diet Versus a Nutritarian Diet

When I started Dr. Fuhrman's program, I weighed 206 pounds and had had diabetes for seven years. His nutritional program enabled me to lose sixty-three pounds and get rid of my diabetes, high blood pressure, and high cholesterol without medication. My LDL went from 168 to 73 in five months (without drugs), and I am now maintaining my healthy weight of 143 pounds.

The most amazing thing is that my ophthalmologist had told me I required laser surgery to treat diabetic retinopathy. When I went back to see him again three months later, after eating the high-nutrient diet, he canceled the surgery because he found that damage to my eyes was no longer there.

I am extremely grateful because I know this life-saving information has added many quality years to my life.

—Martin Milford

More than 85 percent of the SAD consists of foods from low-nutrient, high-calorie processed foods, animal products, dairy products, and sweets. These all contribute to excessive weight, high cholesterol, and high blood pressure, so it's no wonder we have an epidemic of diabetes. Natural plants such as vegetables and beans contain thousands of protective micronutrients, such as antioxidants and phytochemicals. When we eat a diet rich in colorful plant foods, we glean a full symphony of nutritional factors that enable better cell function and resistance to aging and stress.

What happens when we combine high-calorie foods without sufficient amounts of protective micronutrients? Cells become congested with waste products such as free radicals and advanced glycation end products (AGEs). The buildup of free radicals and AGEs in cells is sometimes called oxidative stress. It can lead to inflammation, cell damage, and premature cell death. AGEs are the critical toxins that cause nerve damage, blindness, and other complications of diabetes. They build up faster in people who eat low-nutrient junk food and also in diabetics with elevated glucose levels.[1]

When we gain weight, we not only produce more damaging toxic waste in our cells, but we also dilute our body stores of nutrients, lowering the micronutrient concentration in our cells. The simple key to a long, disease-free life is to weigh less and keep a high level of micronutrients in our cells. We need to be relatively thin but well-nourished with micronutrients.

The American diet contains very little nutrient-rich food. Overall, Americans consume 62 percent of their calories in processed foods and 25.5 percent from animal products. This is the crux of the problem. Both processed foods and animal products are deficient in antioxidants and phytochemicals.

We could not design a better plan to prematurely kill off our population. Only 10 percent of American food intake is from vegetables, beans, seeds, nuts, and fruits—the natural high-micronutrient foods that help prevent and reverse diabetes.

The secret: eating more nutrient bang
for each caloric buck.

U.S. Food Consumption by Calories

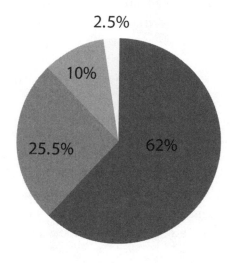

2.5%

10%

25.5%

62%

■ Processed foods: Oil, Sweets, Refined Grains

■ Meats, Eggs, Dairy, Fish

▨ Unrefined Plant Food: Fruit, Vegetables, Beans, Nuts, Seeds

☐ Whole Grains

USDA Economics Research Service, 2005; www.ers.usda.gov/publications/EIB33; www
.ers.usda.gov/Data/FoodConsumption/FoodGuideIndex.htm#calories

Much is known about nutrition and its power to create disease or protect against disease. But the unanswered questions for the majority of our population are: What constitutes a healthy diet? How do we know if our chosen diet is disease producing or disease protecting? What degree of dietary excellence makes a diet reverse a disease?

A nutritarian diet is not just about weight loss. It answers these issues with logic, math, and science. It gives individuals the ability to measure and intuitively judge the nutritional quality of their diets and discern whether it is adequate. What constitutes a healthy diet for a healthy person with good family history and no health problems? How do we design the right diet for those with multiple risk factors or a poor family history? What about people who are faced with serious health challenges? How should their diets be structured for maximum therapeutic effects?

A Nutrient Breakdown

There are two kinds of nutrients: macronutrients and micronutrients. Here's a simple definition of each:

- Macronutrients are nutrients that supply the calories our bodies need for energy and growth.

- Micronutrients are nutrients that appear in trace amounts in foods but are essential for health and growth and that do not contain calories.

There are four macronutrients in the foods we eat: water, carbohydrates, proteins, and fat. Because water is calorie-free, we will not consider it now. All the foods we eat contain some combination of the three calorie-containing macronutrients. If you eat too many macronutrients, you are overeating calories, which causes weight gain, chronic conditions, and premature death.

Yes, to lose weight and improve your health, you need to eat less fat, less carbohydrate, and less protein, reducing total caloric intake. But the secret is not to count calories to reduce calories. That never works. The secret is to focus on micronutrients. I know it defies logic, at first, but true health lies in a high-quality diet—eating foods packed with micronutrients.

Micronutrients are where the magic happens. These nutritional substances in the foods we eat don't contain calories, but they *do* contain the very nutrients that heal the body. Micronutrients are needed for your body to rid itself of waste, repair damage, and support normal day-to-day functions.

Micronutrients include fourteen vitamins and sixteen essential minerals known to be vital to human health, and the importance of incorporating them into your diet for overall health cannot be overstated. Their impact on health is broad and vast. However, these vitamins and essential minerals, identified over seventy-five years ago, are just two types of micronutrients.

Phytochemicals are the third type of micronutrient and were identified more recently. The various kinds of phytochemicals are still being discovered, and a comprehensive list of their many functions has yet to be completed. In the last decade, we found that foods contain thousands of beneficial micronutrients in addition to the original vitamins and minerals discovered back in the 1940s. Now we know the major micronutrient load in food is not vitamins, not minerals, but phytochemicals. These substances pack a powerful punch. They function to improve human health and longevity. We found the fountain of youth, and it was right in front of our noses all along. There are tens of thousands of phytochemicals in natural, whole, vegetable-based foods. These plant nutrients are essential in helping protect you from disease. If you are already sick, they can help you recover.

All of these life-protecting and life-saving nutrients are found in whole foods. Vegetables, beans, berries, and seeds are particularly high in these nutrients. They are the key to optimal health as well as disease reversal and protection.

Remember my health equation:

$$H = N/C$$

This means your health is dependent on the nutrient-per-calorie density of your diet. The quality of a diet can be judged based on three simple criteria:

1. Its level of micronutrients (vitamins, minerals, and phytochemicals) per calorie

2. Adequate macronutrients (fat, carbohydrates, and protein) to meet individual needs but without excess calories that may lead to overweight or health compromise

3. Avoidance of potentially toxic substances (such as trans fats) or substances harmful in excess (such as sodium)

My health equation $H = N/C$ expresses this simple concept of eating for micronutrient per calorie density. The foods with the highest micronutrient density have the most powerful therapeutic effect and are the most effective in promoting weight loss and reversing diabetes.

ANDI

The concept of micronutrient density is put into action by looking at an assortment of foods and analyzing the micronutrients they contain. I have ranked the nutrient density of many common foods in the table on page 49 using my Aggregate Nutrient Density Index (ANDI).* This index assigns scores to a variety of foods based on how many nutrients they deliver to your body in each calorie consumed. Each of the food scores is out of a possible 1,000 based on the nutrients-per-calorie equation.

---

*To determine the ANDI scores, an equal-calorie serving of each food was evaluated. The following nutrients were included in the evaluation: fiber, calcium, iron, magnesium, phosphorus, potassium, zinc, copper, manganese, selenium, vitamin A, beta carotene, alpha carotene, lycopene, lutein and zeaxanthin, vitamin E, vitamin C, thiamin, riboflavin, niacin, pantothenic acid, vitamin B_6, folate, vitamin B_{12}, choline, vitamin K, phytosterols, and glucosinolates plus the ORAC score. ORAC (Oxygen Radical Absorbance Capacity) is a measure of the antioxidant, or radical, scavenging capacity of a food. For consistency, nutrient quantities were converted from their typical measurement conventions (mg, mcg, IU) to a percentage of their Reference Daily Intake (RDI). For nutrients that have no RDI, goals were established based on available research and current understanding of the benefits of these factors. Points were added if the food item was antiangiogenic or contained organosulfides, aromatase inhibitors, resistant starch, or resveratrol.

To make it easier to compare foods, the raw point totals were converted (multiplied by the same number) so that the highest ranking foods (leafy green vegetables such as mustard greens, kale, and collards) received a score of 1,000, and the other foods received lower scores accordingly.

Because nutritional labels don't give you the information necessary to understand exactly what you are eating, these rankings do the equation for you and give you a sense of what foods score the highest. You can use this index to estimate the quality of your current diet or to plan for an improved diet. Using ANDI is simple—it is meant to encourage you to eat more foods that have high numbers and to eat larger amounts of these foods. The higher the number and the greater percentage of those foods in your diet, the better your health.

Because phytochemicals are largely unnamed and unmeasured, these rankings underestimate the healthful properties of colorful natural plant foods compared to processed foods and animal products. One thing we do know is that the foods that contain the highest amount of known nutrients are the same foods that contain the most unknown nutrients too. So even though these rankings may not consider the phytochemical number sufficiently, they are still a reasonable measurement of their content.

Vegetables clearly walk away with the gold medal—no other food is even close. So, of course, green vegetables have the best association with lower rates of cancer and heart disease. While the majority of most people's caloric intake is from the lower end of this table, people who move their consumption higher will substantially protect their health. And the recipes and meal plans in this book will help you reach this goal.

When you seek to consume a broad array of both discovered and undiscovered micronutrients via your food choices, you are a nutritarian. It is not sufficient to merely avoid trans fats or saturated fats. It is not sufficient for the diet to have a low glycemic index. It is not sufficient for the diet to be low in animal products. It is not sufficient for the diet to be mostly raw food. A truly healthy diet must be micronutrient rich, and the micronutrient richness must be adjusted to meet individual needs. Because the foods with the highest micronutrient-per-calorie scores are green vegetables, beans, colorful vegetables, berries, and other fruit, the consumption of enough of these foods is required to meet our micronutrient needs and to promote reversal of diabetes. Not only is it necessary to ingest a high enough absolute value of micronutrients, but the full breadth of micronutrient diversity is also needed for superior health. I call this comprehensive micronutrient adequacy CMA.

In diabetes research, the glycemic index (GI) of carbohydrates has long been recognized as a favorable aid for diabetics to control blood sugar. The same is now often the case in lipid research, as it has been demonstrated that high glycemic diets, rich in white flour, refined sweets, and processed foods are unfavorable to both glucose levels and lipid parameters.[2] The GI is a ranking of carbohydrates on a scale from 0 to 100 according to the extent to which they raise blood sugar levels after eating. Foods with a high GI are those which are rapidly digested and absorbed and result in marked fluctuations in blood sugar levels. Low-GI foods, by virtue of their slow digestion and absorption, produce gradual rises in blood sugar and insulin levels, and they have proven benefits for health. Refined foods made from sugar and white flour are not only high-glycemic foods, but they are also nutritionally deficient and induce micronutrient loss. The glycemic load (GL) is the number that considers the glycemic index within a given serving of food and the actual calories of glucose produced, thereby making it more practical for calculating the overall blood sugar–raising effect of a serving, meal, or daily menu.

Those who advocate a high-protein (meat-based) diet, hang their hat on the low GI of animal products to explain the advantages of a diet rich in animal products and lower in vegetation. This view oversimplifies the multifactorial nuances of nutrition and results in a distorted understanding of nutritional science.

Ranking food on GI alone ignores many other factors that may make that food favorable or unfavorable. Because a carrot has a higher GI than a slice of bacon does not make the bacon a better food for a

DR. FUHRMAN'S ANDI SCORES

Kale	1000	Tofu	82
Watercress	1000	Beans (all varieties)	71
Collards	1000	Seeds: Flax, Sunflower, Sesame, Hemp, Chia (avg)	68
Bok Choy	865	Green Peas	63
Spinach	707	Cherries	55
Arugula	604	Apples	53
Romaine Lettuce	510	Peanut Butter	51
Brussels Sprouts	490	Corn	45
Carrots/Carrot Juice	458	Pistachios	37
Cabbage	434	Oatmeal	36
Broccoli	340	Salmon	34
Cauliflower	315	Milk, 1%	31
Mushrooms	238	Eggs	31
Red Bell Peppers	265	Bananas	30
Asparagus	205	Walnuts	30
Tomatoes	186	Whole Wheat Bread	30
Strawberries	182	Almonds	28
Blackberries	171	Avocados	28
Leeks	135	White Potatoes	28
Raspberries	133	Cashews	27
Blueberries	132	Chicken Breast	24
Iceberg Lettuce	127	Ground Beef (85% lean)	21
Pomegranates/Pomegranate Juice	119	White Bread	17
Grapes	119	White Pasta	16
Cantaloupe	118	Low-fat Cheddar Cheese	11
Onions	109	Olive Oil	10
Plums	106	Corn Chips	7
Oranges	98	Cola	1
Cucumbers	87		

diabetic or heart patient. There are other important nutritional consid-
erations besides GI, including the toxicity, micronutrient density, and
fiber. Good examples of such nutritional nonsense include Dr. Barry
Sears of the Zone Diet, who warns against the consumption of lima
beans, papayas, and carrots because of their GI; and Dr. Robert Atkins,
who excluded fruits and vegetables with powerful anticancer benefits
from his diet.

Glycemic Index and Glycemic Load of Common Carbohydrate-Containing Foods[3]

Carrots are a good example of the lack of precision inherent in using
only the glycemic index. They are high in fiber and nutrient rich, but
their GI is 35. Carrots are relatively low in calories, and when they are
eaten raw their glycemic effect is lessened further, as the body does not
absorb all of the calories in raw foods. The GL is the accurate measurement here, not the GI. Carrots are not a negative food, even for the
diabetic, as the GL is only 3. This is why raw carrots are a favorable
weight loss–promoting food. Instead of focusing narrowly on the concept of GI, we have to consider the other values of the food as well as
the healthful qualities and GL of the entire meal when put together. By
the way, weight loss and micronutrient adequacy are more important
than minor and temporary fluctuations in blood sugar, because they
lead to long-term wellness and resolution of the diabetic condition.

Studies evaluating the negative effects of a higher glycemic diet re-
vealed that foods composed of low-nutrient, low-fiber, processed grains
and sweets have deficiencies, and they harm far beyond their glycemic
response. Processed foods are also low in fiber, phytonutrients, and antioxidants and are rich in toxic acrylamides. In addition to having a high
GL, they are disease-promoting foods. When a diet is rich in nutrients,
the disease-protective qualities of these foods and their weight-loss benefits overwhelm any insignificant drawback from their moderate GL.

UNDERSTANDING THE GLYCEMIC INDEX

FOOD	GLYCEMIC INDEX	GLYCEMIC LOAD
White Potato (1 medium baked)	90	29
White Rice (1 cup cooked)	68	29
Brown Rice (1 cup cooked)	58	24
White Pasta (1 cup cooked)	53	21
Chocolate Cake (1/10 box cake mix + 2T frosting)	38	20
Raisins (1/4 cup)	64	19
Corn (1 cup cooked)	52	18
Sweet Potato (1 medium baked)	69	14
Black Rice (1 cup cooked)	65	14
Grapes (1 cup)	59	14
Rolled Oats (1 cup cooked)	55	13
Whole Wheat (1 cup cooked)	30	11
Mango (1 cup)	51	11
Lentils (1 cup cooked)	40	9
Apple (1 medium)	39	9
Kiwi (2 medium)	58	8
Green Peas (1 cup cooked)	53	8
Butternut Squash (1 cup cooked)	51	8
Kidney Beans (1 cup cooked)	22	7
Blueberries (1 cup)	53	7
Black Beans (1 cup cooked)	20	6
Watermelon (1 cup)	76	6
Orange (1 medium)	37	4
Carrots (1 cup cooked)	39	3
Carrots (1 cup raw)	35	2
Cashews (1 ounce)	25	2
Strawberries (1 cup)	10	1
Cauliflower	negligible	negligible
Eggplant	negligible	negligible
Tomatoes	negligible	negligible
Mushrooms	negligible	negligible
Onions	negligible	negligible

Recently a systematic review was performed of published human intervention studies comparing high- and low-GI foods or diets and their effects on appetite, food intake, energy expenditure, and body weight. In a total of thirty-one short-term studies, the conclusion was that there is no evidence that low-GI foods are superior to high-GI foods in regard to long-term body weight control.[4] More recent research compared the exact same caloric diets, one with a lower and one with a higher GL, and demonstrated that lowering the GL and GI of weight-reduction diets does not provide any added benefit to calorie restriction in promoting weight loss in obese subjects.[5] So the GI and GL are important, but they cannot be the primary focus of a healthy diet. They are just one of many aspects to be considered when understanding what makes this proposed diet style ideal. This will come into play in the design of the optimal diet and best carbohydrate choices in chapter 6.

The important point to remember is that a diet with a high micronutrient density already has a favorable GL. It is also low in saturated fat, high in fiber, rich in phytochemicals, and naturally alkaline. In other words, instead of focusing on one positive aspect alone, consider

Nutritarian Food Pyramid for Diabetics

all the positive features of what makes a diet style disease protective. Fad diets too often rely on one aspect of food and digestion regardless of the potential positive and negative factors that exist simultaneously. So the GL plays a role in designing the optimal reversal diet for a diabetic, but let's not allow the GI or the GL to be the sole determinant of our diet.

Also keep in mind that nutrient-density scoring is not the only factor that determines good health. For example, if we ate only foods with a high nutrient-density score, our diets would be too low in fat. So we have to pick some foods with lower nutrient-density scores (but preferably the ones with the healthier, higher nutrient-containing fats such as seeds and nuts) to include in our high-nutrient diet. Additionally, if thin or highly physically active people ate only the highest-nutrient foods, they would become so full from all of the fiber and nutrients that they would be unable to meet their caloric needs and would eventually become too thin. This, of course, gives you a hint at the secret to establishing a permanent low body-fat percentage if you have a metabolic hindrance to weight loss. But, shhh, don't tell anybody about this.

Optimal health cannot be expected without attention to the consumption of high-micronutrient foods. For example, a vegan diet, centered on high-starch foods such as white rice, white potatoes, refined cereal grains, and bread products, does not contain sufficient micronutrient richness for maximizing longevity. In some susceptible individuals, the lack of attention to micronutrient density may even be disease causing.

Hundreds of individuals have lost over a hundred pounds, some even more than two hundred pounds, and several more than three hundred pounds by following this nutritarian diet. Countless others have just lost the amount of weight they needed to earn back their health. But it is not just about weight loss. Utilizing large volumes of nutrient-rich vegetation in the diet has been demonstrated to lower cholesterol more effectively than cholesterol-lowering drugs.[6] My patients routinely and predictably see their blood pressure return to normal and their atherosclerotic heart disease or peripheral vascular disease melt away as well.

Another revolutionary finding besides the importance of consuming a sufficient quantity and variety of nutrients is that high-nutrient eating suppresses your appetite. You naturally desire fewer calories. So although this book is about eating less, you don't realize you are eating fewer calories and you don't desire more calories. The nutritarian diet style blunts your desire to overeat. In the following pages, we will discuss this added benefit and the ins and outs of hunger and cravings.

CHAPTER FOUR

Reversing Diabetes Is All About Understanding Hunger

Dr. Glen Paulson was a forty-year-old chiropractor and father of four. He suffered from uncontrolled type 2 diabetes, diabetic neuropathy, kidney stones, high cholesterol, and obstructive sleep apnea. He weighed 330 pounds, his fasting blood glucose level was 240, his HbA1C level was 10.4, and his blood pressure was 145/90 on metformin 1,000 milligrams twice daily and Glyberide 5 milligrams twice daily. His physician wanted him to go on insulin because his blood sugar could not be controlled on oral medication and also wanted him to add more medication to further lower his high triglycerides and high blood pressure.

Dr. Paulson recalls, "My kidneys were shutting down, I had stones, and I was constantly in pain. My doctor told me if I didn't change my diet, I would need dialysis in a few years. When I turned down the medication request the nurse gave me over the phone, the doctor called me back and explained the risks to my health and how serious a matter it was. I got off the phone, and I just cried.

"I read Eat to Live *and decided to change. I had been ignorant and reckless with my health." Eight weeks later, when*

Dr. Paulson went back for a checkup, his physician hugged him and said he never saw anyone reverse so many health problems just from diet and exercise.

After six months of following my advice, Dr. Paulson lost eighty pounds, his fasting blood glucose level lowered to 90, his HbA1C level went to 6.5, and his blood pressure reduced to 120/70. His only medication at the six-month marker was metformin 1,000 milligrams twice daily.*

Dr. Paulson's wife, Jillian, also lost thirty pounds. She told us Glen "is doing so much better. He has been a good example for his patients, and they are changing their diets as well. When they see what happened to Glen, they all want to lose weight and get off their medications too because of his good example. This is the best lifestyle change we have ever made, and we 100 percent promote this plan. We now teach a health class on this once a month, and it has been phenomenal. Thanks for everything."

A high-micronutrient diet does not just improve health for your body, but it also decreases food cravings and sensations leading to overeating behavior. Individuals adopting a diet style rich in micronutrients report a change in the perception of hunger signals. The sensations commonly considered hunger, and even reported in medical textbooks as such, appear to dissipate for the majority of people, and a new sensation that I label true or throat hunger arises instead.

A diet too low in micronutrients leads to heightened oxidative stress. Oxidative stress means inflammation in the cells due to excessive free radical activity. It is accompanied by a buildup of toxic metabolites that can create physical symptoms of withdrawal when digestion ceases in between meals. Besides the toxins we consume from food, cells pro-

---

*If he was my patient I would have also lowered the metformin dose by this point, or perhaps discontinued it.

duce their own metabolic wastes that need to be removed from cells and tissues.

When our diets are low in phytochemicals and other micronutrients, we build up intracellular waste products. It is well accepted in scientific literature that toxins such as free radicals, AGEs, lipofuscin, and lipid A2E build up in tissues when people's diets are low in micronutrients and phytochemicals, and that these substances contribute to disease.[1]

It has already been noted that overweight individuals build up more inflammatory markers and oxidative stress when fed a low-nutrient meal compared to normal-weight individuals.[2] Because of this, people prone to obesity experience more withdrawal symptoms that direct them to the overconsumption of calories. These are the sources of the toxic hunger cravings that often lead to binging and other gut-busting behavior. It is a vicious cycle promoting the problem and preventing its resolution. Those with healthier diets do not build up such high levels of inflammatory markers and as a result do not experience intense withdrawal hunger symptoms.[3]

Phytonutrients are required for the body to properly detoxify metabolic waste products—they enable cellular detoxification. When we don't eat sufficient phytochemical-rich-vegetation and instead consume low-nutrient food and excess animal proteins (creating excess nitrogenous wastes) we often exacerbate the buildup of metabolic waste products in our bodies.[4] These wastes are just like drug toxins.

The withdrawal symptoms, conventionally called hunger, develop from inadequate or poor nutrition. I call these withdrawal symptoms toxic hunger. It is important for us to understand and differentiate toxic hunger from true hunger. Toxic hunger appears at the lower plateau of the blood sugar curve, drives overeating behavior, and strongly increases the desire to consume more calories than the body requires, leading to weight gain and diabetes. True hunger, however, appears when the body has used up most of the calories from the previous meal as well as the stored glucose (stored as glycogen) and is ready to be refueled. With a change of diet, toxic hunger gradually lessens and resolves, allowing individuals to be satisfied eating less.

When you adopt this nutritarian diet, becoming healthy is the first step. You soon find that the symptoms of toxic hunger are gone. Instead, you will eventually experience the feeling of true hunger, which encourages the precise amount of calories required for good health and the maintenance of ideal weight. True hunger serves as an important guide to promote enjoyment of food. It gives us precise signals from our bodies so we know the amount of calories needed to sustain our lean body mass. When we eat when we are hungry, food tastes much better and we are physiologically primed for proper digestion. Hunger, in the true sense of the word, indicates that it is time to eat again.

TYPICAL SYMPTOMS OF TOXIC HUNGER

Feeling of emptiness in stomach

Gurgling, rumbling in stomach

Dizziness or lightheadedness

Headache

Irritability or agitation

Lack of concentration

Nausea

Shakiness

Weakness or fatigue

Impairment in psychomotor, vigilance, and cognitive performances

TYPICAL SYMPTOMS OF TRUE HUNGER

Throat and upper chest sensation

Enhanced taste sensation

Increased salivation

The critical message is that the wrong food choices lead to withdrawal symptoms that are mistaken for hunger. You can always tell that these are toxic hunger symptoms because you experience shakiness, headaches, weakness, and abdominal cramps or spasms. Initially, these symptoms are relieved after eating, but the cycle simply starts over again with the symptoms returning in a matter of hours. Eating when you experience toxic hunger is not the answer. Changing what you eat to stop toxic hunger is.

When our bodies become acclimated to noxious or toxic agents, it is called addiction. If we try to stop taking nicotine or caffeine, we feel ill. This is called withdrawal. When we stop doing something harmful to ourselves, we feel ill because the body attempts to mobilize cellular wastes and attempts to repair the damage caused by the exposure. If we drink three cups of coffee a day, we would get a withdrawal headache when our caffeine level dipped too low. When we consume more caffeine again, we feel a little better because it retards detoxification, or withdrawal. In other words, the caffeine withdrawal symptoms can contribute to our drinking more caffeine products.

Similarly, toxic hunger is heightened by the consumption of caffeinated beverages, soft drinks, and processed foods. Toxic hunger appears after a meal is digested and the digestive track is empty, and it can feel extremely uncomfortable, which can make us think we need to eat or drink a caloric load for relief.

The confusion about food-addictive behavior is compounded because when we eat the same heavy or unhealthful foods that are causing the problem to begin with, we initially feel better. This makes becoming overweight inevitable, because if we stop digesting food, even for a short time, our bodies will begin to experience symptoms of detoxification or withdrawal from our unhealthful diet. To counter this, we eat heavy meals, eat too often, and keep our digestive track overfed to lessen the discomfort from our stressful diet style. In other words, we keep eating too often and too much to postpone or mitigate the physical discomfort caused by our bad diet.

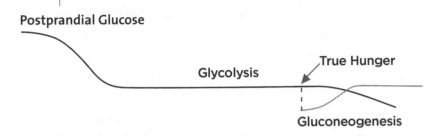

The glucose absorbed right after a meal is called postprandial glucose. After the carbohydrates from the meal are broken down to simple sugars and eventually utilized or stored in the body, most of the glucose not burned is stored as glycogen in the liver and muscle tissues. Glucose is continually utilized to fuel our cells and especially our brain. Our brain use makes up 80 percent of our caloric needs in the resting state. After the meal's contribution is utilized and digestion ceases, we start to gradually burn down our candle of stored glycogen in the liver as our glucose source. This catabolic or breakdown phase, when stored glycogen is our main source of glucose, is called glycolysis. When glycogen stores are being burned for glucose, toxins are better mobilized for removal and repair activities are heightened. Spending time in glycolysis, while resting the digestive apparatus in this non-feeding stage, is important for health and a long life.

But Americans, and especially diabetics, become uncomfortable when beginning glycolysis. They don't feel right if they delay eating too long. This is an important reason why they became diabetic to begin with. They must overeat to feel okay. Just like a person addicted to tobacco must smoke cigarettes just to feel okay, they have become addicted to their dangerous and toxic diet habits and they can't tolerate the symptomatic detoxification events that occur during glycolysis.

Most often these uncomfortable symptoms occur simultaneous to our blood sugar decreasing and glycolysis beginning, but they are not caused by hypoglycemia. While we feed off glycogen stores, rather than actively digest and assimilate glucose, our bodies cycle into heightened detoxification activity—so these sick feelings that accompany glycoly-sis are a result of tissue sensitivity to mobilization of waste products, which occurs when most active digestion is finished. They occur when the blood sugar is at its lower plateau. These symptoms are obviously not merely caused by low blood sugar, though the symptoms occur in parallel with lower blood sugar.

Gluconeogenesis is the breakdown of muscle tissue to fabricate glu-cose after glycogen stores have been depleted. As the liver's glycogen stores are utilized and diminish, true hunger signals the need for calo-ries before muscle breakdown begins, thus preventing the onset of glu-coneogenesis. Gluconeogenesis becomes activated after the glycogen stores have been depleted, so if fasting is continued too long, the body would utilize muscle tissue as a glucose source. Does the body want to waste muscle to maintain our glucose levels? Of course not. We get a clear signal to eat before that begins. I call this clear signal true hunger. True hunger is protective of our muscle mass and gives a clear signal to eat before the beginning of gluconeogenesis, as the glycogen stores are running low and glycolysis is winding down.

Phytonutrients are required for the body to properly detoxify meta-bolic waste products as they enable cellular detoxification. Oxidative stress is caused by an imbalance between the production of reactive oxygen and a biological system's ability to readily detoxify the reac-tive intermediates or easily repair the resulting damage. This oxidative stress from the buildup of toxins leads to diseases, including most of the conditions commonly considered the complications of diabetes.

All forms of life maintain a reducing environment within their cells. That means they are continually removing wastes and remov-ing free radicals. Disturbances in this normal redox state occur from micronutrient deficiencies and can cause toxic effects through the pro-duction of peroxides and free radicals, which damage all components of the cell. When oxidative stress occurs, certain by-products are left behind and are excreted by the body, mostly in the urine. These by-products are oxidized DNA bases, lipid peroxides, and malondialdehyde from damaged lipids and proteins. The higher the levels of these various markers (which can be measured in the urine), the greater the damage to the cells—marking the advancement of an oxidative stress-induced disease.

The mammalian circadian system is organized in the brain's hypothalamus. This section of the brain synchronizes cellular oscillators in most peripheral body cells. The liver glucose sensor activates these parts of the brain involved in cellular cycles. Fasting-feeding cycles accompanying rest-activity rhythms are the major timing cues in the synchronization of most peripheral clocks, especially metabolic activity and cellular detoxification. Detoxification efforts of the body vary cyclically and correspond with the rhythm and repetitive timing of sleeping and eating. The deactivation of noxious food components by hepatic, intestinal, and renal detoxification systems is among the metabolic processes regulated in a cyclic manner. The detoxification output of by-products is enhanced during cyclic periods corresponding with glycolysis.[5] This means that when we are not digesting food, the body is in an enhanced repair and detoxification cycle.

Not only does eating low-nutrient food build up more toxins, creating inflammation and disease in our cells and organs, but the buildup of these toxins also leads us to feel ill the minute our digestive tract is no longer busy digesting. So we are almost forced to overeat to prevent withdrawal symptoms. As soon as the digestion of food is complete, changes in all body systems begin to occur. Some patients report symptoms even within a few hours of not eating a food. Coffee drinkers, for example, are usually on an obligatory ingestion cycle and may get withdrawal headaches and cravings within hours of missing regular coffee doses.

It has already been noted that overweight individuals build up more toxic waste products and express heightened inflammatory markers and oxidative stress when on a low-nutrient meal compared to normal-weight people.[6] Because of this, men and women prone to obesity experience more withdrawal symptoms, causing overconsumption of calories. It is a vicious cycle promoting the problem and preventing its resolution.

People who eat healthier diets do not build up inflammatory markers nearly as much as people who don't.[7] The point here is that many people are overweight because their body is forcing them to require more calories just to feel normal. They don't feel well when they attempt to eat the amount of calories better aligned with their metabolic needs. A critical feature that makes a person overweight or diabetic is the environment of excessive and toxic food in the modern world. This in turn leads to the chronic oxidative and inflammatory stress created by modern food choices. This cellular "dis-ease" then creates symptoms that support its continuation, just like a cocaine addict seeks cocaine. This is why diets always fail. The secret to beating this vicious cycle is to focus on micronutrient quality. Only then will the desire for excessive calories cease.

Every cell is like a little factory—it makes products; produces waste; and then must compact, detoxify, and remove the waste. If we let waste metabolites build up through the consumption of processed grains, oils, and animal products and insufficient consumption of vegetation, the body will attempt to mobilize these wastes (creating discomfort) when it can. But it only can do that effectively if it's not actively digesting food. Eating alleviates the discomfort because it halts or delays the detoxification process.

What I have observed and quantified with not merely hundreds but with thousands of individuals is that the drive to overconsume calories is blunted by high-micronutrient, high-antioxidant food consumption. The symptoms that people thought were hypoglycemia or even hunger simply disappear after eating healthfully for a few months. After a two- to four-month window, when micronutrients in the body's tissues are enhanced, people not only lose symptoms of fatigue, headaches, irritability, and stomach cramping, but they also get back in touch with true hunger felt in the throat. This sensation, they report, makes eating more pleasurable and better directs them to an appropriate amount of calories for their body's biological needs.

My discovery documenting the changing perception of hunger, resulting in a lower caloric drive in more than seven hundred people eating a high-nutrient diet, was published in *Nutrition Journal* in November 2010.[8] Of interest was that hunger became a sensation in the upper chest and throat for 90 percent of those compliant with the dietary recommendations and also that it took three to six months for most of the participants to experience this change in hunger and the lessening of hunger symptoms. Three to six months corresponds with the time frame it takes to achieve adequate tissue levels of phytochemicals after dietary excellence is begun. The study's conclusion was:

> *Our findings suggest that it is not simply the caloric content, but more importantly, the micronutrient density of a diet that influences the experience of hunger. It appears that a high nutrient-density diet, after an initial phase of adjustment during which a person experiences "toxic hunger" due to withdrawal from pro-inflammatory foods, can result in a sustainable eating pattern that leads to weight loss and improved health. A high nutrient-density diet provides benefits for long-term health as well as weight loss.*

True hunger signals when our bodies need calories to maintain our lean body mass. When we eat food demanded by true hunger and true hunger only, we do not become overweight to begin with. However, in our present toxic food environment, we have lost the ability to connect with the body signals that tell us how much food we actually need. We have become slaves to withdrawal symptoms and eat all day long, even when there is no biological need for calories. The body has a compounded sensation of hunger and cravings that, for most, is simply overwhelming. As a result, people are either unable to lose weight or are unable to keep it off.

In an environment of healthy food choices, we would not feel any symptoms after a meal until our bodies actually require more nourishment. Our bodies have the beautifully orchestrated ability to give us the precise signals that tell us exactly how much to eat to maintain an ideal weight for long-term health and well-being. Thousands of people have learned this and have demonstrated that this phenomenon is real. After learning and applying this information, many have lost over a hundred pounds, some more than three hundred pounds, without surgical intervention and have kept the weight off.

In a portion-controlled (calorie-counting) diet, it is likely that the body will not consume adequate fiber or nutrients. The body will have a compounded sensation of hunger and craving that, for most, is impossible to control. Invariably, it results in failure to lose weight or the cycle of losing weight and eventually gaining it back. Calorie counting simply doesn't work in the long run. Diets based on portion control and calorie counting generally permit the eating of highly toxic, low-nutrient foods and then require us to fight our addictive drives and attempt to eat less. This combination undernourishes the body, resulting in uncontrollable and frequent food cravings. Without an adequate education in nutrition and solid principles to stick to, people on these diets are forced to flounder and fail, bouncing from one diet to another, always losing a little and regaining it. They frequently regain more than they lost.

Life is prolonged by eating less while maintaining a high-nutrient cellular environment. However, trying to eat fewer calories is ineffective and almost futile. The secret is to desire fewer calories. The high consumption of low-calorie, high-nutrient foods such as raw vegetables, cooked greens, beans, and seeds prepared in delicious combinations makes you feel physically full from all the fiber and satisfied from all the chewing. You lose the addictive cravings, and then you simply and naturally desire less food. It makes it quite simple to lose lots of weight. The core elements to this revolutionary diet are:

- Micronutrient-per-calorie density is important in devising and recommending menu plans and dietary suggestions for the most effective approach for both weight loss and for preventing and reversing diabetes and heart disease.

- Low-nutrient eating (and toxic eating) leads to increasing cellular toxicity with undesirable levels of free radicals and AGEs. This toxicity causes addictive withdrawal symptoms (toxic hunger), which result in more frequent eating and overeating.

- Dietary micronutrient quality (H=N/C) must be increased accordingly to utilize dietary recommendations therapeutically for disease reversal or to protect high-risk individuals.

This is not a diet in the sense of something you do to lose weight. This is a new diet style for life—a diet style that every American has the right to know about, so we all have the choice to protect our precious health. It is healthful eating. It is effective for long-term weight control because it modifies and diminishes the sensations of so-called hunger, enabling overweight individuals to be more comfortable eating fewer calories. Make the change for life.

CHAPTER FIVE

High-Protein, Low-Carb Counterattack

Jessica was a forty-eight-year-old mother of two teenagers, who originally weighed 193 pounds. She was what I call a vegejunkarian, eating highly processed vegetarian foods and few fresh fruits or vegetables. Looking back on her health at that time, Jessica said, "My health was in the toilet! I had aches and pains all over, including chest pains, severe diabetic [frozen] shoulder that caused my arm to be nearly immobile, and horrible gastroesophageal reflux disease. I napped three times a day. I also had a host of skin conditions, rosacea, and tendonitis in my left hand. I had chronic insomnia and had to get up two or three times a night to go to the bathroom. My vision was really blurry from the out-of-control sugars. I had neuropathy in two toes in my right foot, and I was losing a lot of hair. I was a walking time bomb!"

Jessica's fasting blood glucose level was 282, her HbA1C was 12.2, and her blood pressure was 150/110. She was on no medications because she was in complete denial and wouldn't go to the doctor.

After searching for an answer and finding my program online, Jessica became a member of my online support center and began following a healthy, nutritarian diet. Six months later, she reported her weight dropped to 151 pounds. Her fasting blood glucose level is now 96, her HbA1C is 5.4, and her blood pressure is 110/70. The only medication she takes is 100 micrograms of levothyroxine per day for her low thyroid function. She now describes her health as "so much better! No more aches or pains. No more gastroesophageal reflux. My frozen shoulder is 98 percent healed. My vision is much better, and my painful neuropathy is completely gone (that took three months). I rarely have interrupted sleep anymore, and I never have to get up at night to pee. My tendonitis seems to be gone, my hair is fuller, my eyes are brighter, and my skin is clear. I am a new women, and it has only been six months."

Jessica was delighted to observe the reactions of people she works with. "I went to my office last Friday—I hadn't been there in six months, as I work from home. Several people absolutely did not recognize me. I spent the whole day explaining this program, and I'm sure I sold a few books . . . that day."

Diabetics mostly die of heart attacks. A meat-based diet promotes atherosclerosis, increases the risk of blood clots, and accelerates kidney failure in diabetics. A diet high in animal products and low in vegetables and beans is the formula for a medical disaster. Diabetics need the opposite: a diet high in vegetables and beans and low in animal products.

Some people have bought into the faulty logic that if sugar and refined grains and other high-glycemic foods raise blood sugar and triglycerides, then we should eat more animal products instead of these refined carbohydrates. Unquestionably, sugar, white flour, and other processed grains are unfavorable and must be removed to achieve good health, but to increase animal products at the expense of vegetables, beans, nuts and seeds, and other low-glycemic, nutrient-rich plant foods (which are protein adequate) is not only dangerous but also reduces the potential for the diabetic to recover and get off all medications.

Carbohydrate-restrictive diets that are rich in animal products can offer some short-term improvement in glucose control and can potentially aid weight loss in some people, but because those diets are too rich in animal products (which do not contain phytochemicals or antioxidants), they incur other significant risks such as cancer, heart disease, and kidney disease. The main problems with recommending a diet with a significant amount of animal products for diabetics are that the increased protein intake promotes the progression of diabetic kidney disease, and the animal-source protein and saturated fat intake raise cholesterol and promote heart disease. Even though a protein-dense diet might offer some marginal weight loss benefits compared to a diet with lots of processed carbohydrates, it still does not allow the substantive weight reduction that diabetics really need to rid themselves of the disease.

Emerging evidence also suggests that carbohydrate-restrictive, also called ketogenic, diets "create metabolic derangement conducive to cardiac conduction abnormalities and/or myocardial dysfunction."[1] In other words, it may cause other potentially life-threatening heart problems. Ketogenic diets are the most dangerous; medical literature has shown them to cause cardiomyopathy, a pathological enlargement of the heart that is reversible but only if the diet is stopped in time.[2] Even following a ketogenic diet short term, such as with the induction phase of the Atkins or Dukan diets, is dangerous, and deaths have occurred from cardiac arrhythmias induced from the electrolyte derangement.[3]

Not only are diets very high in animal products dangerous in the short-term, they are more dangerous when followed long term. Animal products need to be restricted for disease reversal to occur predictably. Diabetics have significantly better chances at reversing their disease when they avoid excess animal protein. Scientific studies have demonstrated that a high intake of animal products creates an excess of branched-chain amino acids, which further inhibit insulin function and worsen diabetes control.[4]

This worsening of diabetes from increased animal product consumption was borne out in a recent study in which researchers analyzed the diets of 38,094 participants from the European Prospective Investigation into Cancer and Nutrition study. Researchers found that for every 5 percent of calories consumed from animal protein, the risk of diabetes increased 30 percent.[5] Increased animal protein intake also coincided with increased body mass index, waist circumference, and blood pressure. Vegetable protein was not associated with increased diabetes risk. Quite a few other studies corroborate that diets that contain meat enable or worsen diabetes. Of note is the most recent Adventist Health Study-2, involving more than sixty thousand men and women, which is revealing because when those eating only a small amount of animal products were compared with those eating none, those following the vegan diet were found to have a diabetes prevalence that was approximately one-third that of the nonvegetarians (2.9 percent versus 7.7 percent). The lacto-ovo vegetarians, pesco-vegetarians, and semi-vegetarians had intermediate diabetes prevalence rates of 3.2 percent, 4.8 percent, and 6.1 percent, respectively.[6] Obviously, the best way to reverse diabetes is to avoid diets centered on processed and high-glycemic carbohydrates and animal protein.

Tip: Eat more foods rich in vegetable protein and
less or no foods with animal protein.

I have seen many diabetic patients on physician-recommended high-protein diets develop kidney or heart problems. Numerous people have suffered and died needlessly because of misinformation. I consider this advice malpractice. This issue still exists. Many doctors are still advocating this diet style for diabetics. Advocates for high-animal-protein diets flood bookstores and the Internet because people want to hear they can eat all these rich foods that they desire. People buy into the hype and often don't understand the dangers until it is too late. Some enthusiastically jump on a bandwagon of pseudoscientific claims that support the continuation of their preferred food habits and food addictions.

These enthusiasts would have you brainwashed with saturated fat. To undo the damage, let's review some more of the evidence. A May 2004 *Annals of Internal Medicine* study showed that a third of Atkins dieters suffered a significant increase in LDL cholesterol and virtually none of them achieved a favorable LDL below 100. My nutritarian diet invariably drops cholesterol radically and is the only diet style tested in the medical literature to drop LDL cholesterol as much or more than cholesterol-lowering drugs, as reported in the medical journal *Metabolism*.[7] The goal is to have nonmedicated LDL cholesterol below 100, and that will almost never occur with a meat-based diet.

A landmark study published in 2000 actually measured what was happening to the arteries of people on low-carb, high-protein diets. Utilizing SPECT scans to directly measure blood flow within coronary arteries, the development of heart disease was examined in sixteen people on a vegetarian diet that was high in fruits and vegetables and in ten people on a low-carb, high-animal-protein diet. The results were shocking. Those sticking to the whole foods vegetarian diet showed a reversal in expected heart disease. Their partially clogged arteries literally got cleaned out, and blood flow to their heart through their coronary arteries increased by 40 percent. Those on the high-protein diet exhibited rapid advancement of their heart disease with a 40 percent decrease in blood flow in the heart's blood vessels.[8] Thus, the only study on the high-protein diet to actually measure arterial blood flow showed that this style of eating is exceedingly dangerous.

The main problem with low-carbohydrate, high-protein diets is that the intake of the high-nutrient plant foods that contain protective fiber, antioxidants, and phytochemicals, is lowered while calorie-dense and nutritionally poor animal products is raised. Both of these factors are known to diminish cardiovascular health and increase the risk of stroke and heart attack. This was further documented in a large study published in the *British Medical Journal*. To conduct the study, researchers examined nearly forty-four thousand Swedish women aged thirty to forty-nine years and followed up an average of fifteen years later. During the fifteen-year study period, 1,270 cardiovascular events took place in the 43,396 women (55 percent ischemic heart disease, 23 percent ischemic stroke, 6 percent hemorrhagic stroke, 10 percent subarachnoid hemorrhage, and 6 percent peripheral arterial disease). Researchers found that cardiovascular disease incidence more than doubled in the low-carb, high-protein followers.[9]

Fortunately, the Atkins diet has lost its luster as a result of studies like these, and more doctors are informing their patients about its dangers. Unfortunately, other diets with similar strategies but different names, such as the paleo or Dukan diets, keep popping up and luring individuals into the same disproven and dangerous eating patterns. Many people are lured into these dangerous diets because they cling to any argument that condones their food preferences. Diabetics can't afford to make such mistakes, because these mistakes of judgment could result in dramatically increased suffering and a curtailed life span. The paleo diet uses a distorted view of ancient history to argue that a diet of 50 to 80 percent animal products is the most life span enhancing. (This recommendation is double to triple the average animal product consumption in America today.) Early humans ate many different types of diets in various parts of the world, but what they ate here or there is not even the relevant question. It is how long they lived, and how long present humans will live (in good health) with various diet styles that is more relevant. The answer to this question is clear as the preponderance of evidence is overwhelming today.

If the increased risks of heart attack and cancer aren't enough of an argument, a large study tracking kidney damage showed that a high-protein diet accelerates kidney damage in people with even very mild compromise to their kidneys.[10] Almost 25 percent of people over forty-five, especially those with diabetes or high blood pressure, have some degree of kidney impairment. Although the study did not proceed long enough to detect kidney damage in those with perfectly healthy kidneys, it's important to note that kidney damage is often not detectable at lower levels of damage. Higher levels of damage make it easy to diagnose, but by then it could be too late to reverse, especially in diabetics. In fact, Dr. Knight, the lead researcher in this study concluded, "The potential impact of protein consumption on renal function has important public health implications given the prevalence of high-protein diets and use of protein supplements." It is also well established that lots of meat equals lots of gout and kidney stones.[11]

In a press release titled "American Kidney Fund Warns About Impact of High-Protein Diets on Kidney Health," AKF Chair of Medical Affairs Paul W. Crawford, M.D., said, "We have long suspected that high-protein weight loss diets could have a negative impact on the kidneys, and now we have research to support our suspicions." Dr. Crawford is worried that the strain put on the kidneys could result in irreversible "scarring in the kidneys." Dr. Crawford also discussed the risk that bodybuilders take in eating high-protein diets while building muscle. He noted, "Bodybuilders could be predisposing themselves to chronic kidney disease because hyperfiltration (the strain on the kidneys) can produce scarring in the kidneys, reducing kidney function."

Dr. Crawford concluded, "Chronic kidney disease is not to be taken lightly, and there is no cure for kidney failure. The only treatments are kidney dialysis and kidney transplantation. This research shows that even in healthy athletes, kidney function was impacted and that ought to send a message to anyone who is on a high-protein weight loss diet."[12]

There is a vast amount of scientific literature supporting what constitutes excellent nutrition. I have reviewed over twenty thousand studies that indicate that what we put in our mouth *does* matter and that we *can* prevent disease with a high-nutrient diet. It is important that we all know that we can no longer deny the dangers from a diet style rich in meat and other animal products.

Humans are primates, and all primates eat a diet of predominantly natural vegetation. If they eat animal products, it is a very small percentage of their total caloric intake. Luckily, we have modern science that shows that most common ailments in today's world are the result of wrong nutritional choices arising from misguided nutritional information. Now our knowledge base has taken a giant leap forward and we can eat a diet rich in phytochemicals from a variety of natural plant foods that can afford us the ability to live a long, healthy life, which was not easily obtained by our ancestors.

With millions of high-protein, low-carb enthusiasts around the world grasping at straws to justify eating a diet rich in animal products, I hope this information serves to counter health claims by these people and perhaps saves a few lives or reduces suffering. Keep in mind that there is a similarity here between my recommendations and those of the high-protein, low-carb advocates: the low-nutrient, high-glycemic junk that most Americans consume is dangerous. However, with the nutritarian diet, the low-GI benefit that a high-protein diet offers is still achieved but with an emphasis on very high-fiber vegetables, beans, and nuts, which avoid the disadvantages that come from eating too many animal products. Because the nutritional quality of the entire diet is so high, with so much fiber and so many micronutrients per calorie, the GI of the whole diet is favorable overall, and triglycerides and blood sugar fall dramatically.

The main point here that I cannot overemphasize is the benefit of nutritional excellence. In describing the bad science utilized to promote low-carb diets, let's always frame it with what a healthy diet should look like. When you eat a truly health-supporting diet, you can expect not only a drop in blood pressure, a decrease in cholesterol, and a reversal of heart disease, but also a resolution of headaches, constipation, indigestion, and bad breath. Dietary excellence enables people to reverse diabetes and gradually lose dependence on drugs. You should not only achieve a normal weight without counting calories and dieting, but you can also gain robust health and live a long life free of the fear of heart attacks and strokes.

Understanding the differences in various dietary choices is critical for the health seeker. Longevity and disease prevention are the ultimate goals of dietary changes. Obviously, weight loss is not the only goal. You can lose weight by smoking cigarettes or snorting cocaine. When you settle for second-class nutritional advice, you doom yourself not only to a shorter life but also to a poor quality of life, suffering from medical problems that could have been avoided.

Not a month goes by that I do not see at least one diabetic patient whose health has been damaged by following a high-protein fad diet. It's sad to tell people like this that the diet they chose has caused permanent damage, such as a heart attack or kidney disease.

But I am able to offer them good news too. Because I am a specialist in nutritional medicine and see many overweight diabetic patients every day, I have the experience to assure patients that they can quickly get off their insulin and other drugs and, in most cases, become completely nondiabetic with this program of dietary excellence. They do this without incurring the risk of a diet burdened with a dangerously high amount of animal products. Not only is dietary excellence safe and overall health promoting, but the amount of weight loss achieved and the reversal of diabetes are dramatic—results that could not been achieved with a high-protein diet.

How Much and What Type of Animal Products Are Permitted?

Nutritional excellence does not have to exclude all animal products. But it has to be very rich in high-nutrient plant foods composing well over 85 percent of caloric intake. The minimal amount of animal products in your diet that may still permit optimal health is not a fixed or determined number, so it can be adjusted for individual differences or needs within the guidelines offered here. However, if you have had diabetes a long time or have heart disease or high blood pressure and are significantly overweight, you'll achieve better results with fewer animal products, not more. Most people do fine with two or three small servings of animal products a week, but for some, even this small amount of animal protein can cause their cholesterol to go into the unfavorable range. In this book, because I am designing the optimal diet for reversing diabetes, I recommend a maximum of only one or two (two- to three-ounce) servings of animal products a week. Of course, I emphasize that if you use animal products on a regular basis, the serving size should be small, such as a condiment to flavor a vegetable dish, stew, soup, or salad, not as a caloric contributor to the meal. I recommend one or two servings of fish per week—such as salmon, sardines, squid, flounder, scrod, or trout—or one or two servings of fish plus one small serving of white meat fowl, totaling less than six ounces per week. No other animal products are recommended.

More than two servings of fish per week are associated with significantly higher incidence of type 2 diabetes.[13] Following almost two hundred thousand individuals for fourteen to eighteen years, researchers found that the risk of developing diabetes rose as fish consumption increased, resulting in a 22 percent increased incidence of diabetes when participants ate fish more than five times a week compared to those who ate fish less than once a month. Researchers are unsure exactly why more fish in the diet worsens the risk for diabetes, but whether it's an effect of the fish fat, the concentrated protein, or the toxins like dioxin or mercury found in fish, it is clear that a fish-heavy diet is not appropriate for diabetics or people at risk of developing diabetes. I want to make it clear that there is no significant benefit from using fish in your diet at all. The healthful omega-3 fats can be ingested via a supplement (even a vegan DHA/EPA supplement). The allowance for this small amount of animal product is discussed here as some people are insistent on not going all the way to a vegan diet.

Red meats are to be avoided completely. Studies on diabetics and meat eating indicate a 50 percent higher incidence of heart disease in people with high red meat intake.[14] Researchers believe this is not associated with the higher level of saturated fat in red meat but instead with the heme iron it contains. It is increased consumption of both processed foods and animal products that is linked to increased mortality, diabetes, and heart problems.[15] Large-scale studies of the metabolic syndrome have linked the incidence of high glucose, abdominal fat, high triglycerides, and high blood pressure in Western societies with red meat, processed meat, fried food, refined grains, and diet soda.[16] When multiple dangerous foods are consumed, it creates a deadly combination. The metabolic syndrome is a cluster of cardiovascular disease risk factors associated with increased risk of diabetes and mortality. Studies invariably show that the most protection, prevention, and reversal of, and lower risk of, heart disease occurs when the diet style is high in vegetables, beans, fruits, and nuts and is very low in animal products.[17]

We have to dramatically reduce both processed foods and animal products if we have a significant medical issue and expect the body to recover. Too often diet proponents want to make just one or the other the villain. One of the most interesting studies emerging this year was the negative effect of eggs on diabetes. Researchers found that people consuming seven eggs a week had a 58 percent higher risk of developing diabetes than those who did not eat eggs.[18] Furthermore, egg and dairy intake are also linked to heightened risk of heart failure, up to a significant 23 percent higher risk.[19] Neither eggs nor the frequent intake of dairy are appropriate for diabetics. They worsen glucose control and increase heart disease risk, which is dangerous for diabetics, who already have heightened heart disease risk.

Again, the preferred animal products are small amounts of fish just once a week, or fish once a week and white meat fowl once a week, keeping the total under six ounces per week. That's it, because this is too important to let anything slow your progress and increase any disease risks.

Reviewing the Facts About Eggs and Diabetes

1. The Nurses' Health Study, Health Professionals Follow-up Study, and Physicians' Health Study reported that diabetics who eat more than one egg a day double their cardiovascular disease or death risk compared to diabetics who ate less than one egg per week.[20]

2. A Greek study of diabetics reported a five-fold increase in cardiovascular death risk in those eating one egg a day or more.[21]

3. A recent study evaluating atherosclerotic plaque in the carotid arteries found that subjects eating more than three eggs a week (compared to less than two eggs a week) had significantly more carotid plaque area—even after statistical controls for multiple potentially confounding factors, including serum cholesterol. The data related that someone who had eaten five eggs a week for forty years would have two-thirds the amount of plaque as someone who smoked one pack of cigarettes a day for forty years, other factors being equal. This indicates that eggs may increase atherosclerotic plaque development in ways unrelated to elevating blood cholesterol.[22]

4. Eating five eggs a week or more is also associated with an increased risk of developing type 2 diabetes, not to mention prostate cancer.[23]

The inevitable conclusion of all the data is that eggs are more harmful to cardiovascular health than earlier studies suggested. They are particularly unsafe for populations at risk of diabetes and cardiovascular disease.

Get Protein from Your Vegetables

Most of my patients tell me that the typical question their friends or family members have about this plant-based diet is how you get enough protein with so few animal products. Many people are still tied to the myth that a diet needs animal products to be nutritionally sound. To add to the confusion, diet books and magazine articles promulgate the myth that more protein is favorable for weight loss and carbohydrates are unfavorable.

If you are overweight, you have consumed more calories than you have utilized. Micromanaging the percent of fat, protein, or carbohydrates you eat isn't going to change the amount of calories much. You need to consume less calories. Therefore, almost all overweight individuals need to consume less protein, less fat, and fewer carbohydrates—the sources of calories. Don't worry about not consuming enough. With the exception of anorexics, it is very rare to find an American deficient in fat, protein, or carbohydrates.

Inhabitants of modern Western societies generally consume more macronutrients, especially protein, than needed. Protein is ubiquitous; it is contained in all foods, not only animal products. It is almost impossible to consume too little protein, no matter what you eat, unless the diet is significantly deficient in calories and other nutrients as well. Protein deficiency is not a concern for anyone in the developed world. Americans already get too much protein, and it hurts us. When you eat a diet rich in green vegetables and beans, you are actually on a diet fairly high in protein because they are protein-rich foods. And of course when your diet is mostly plant protein, you get your protein packaged with protective fibers, antioxidants, and phytochemicals—a horse of a different color.

But should we carry around little pocket calculators and track everything we eat to make sure we don't accumulate more than 10 percent of our calories from fat? Do we have to watch what we eat to make sure we get enough protein? The reality is that the precise ratio of these nutrients doesn't matter much. What matters is that you are not deficient in any needed macronutrient, that you are not consuming excess calories or excess of anything else that may be harmful, and most importantly, that you meet all your micronutrient needs without over-consuming calories. Simply put, the goal of a healthy diet is to get the most micronutrients, both in amount and diversity, from the fewest calories. And fewer calories means less protein too. The real concern should be getting too much protein, not too little.

The focus on the importance of protein in the diet is one of the major reasons the American public has been led down the path to dietary suicide. We have equated protein with good nutrition and tend to believe that animal products, not vegetables and beans, are the most favorable source of protein. We bought a false bill of goods, and the dairy- and meat-heavy diet has brought forth an epidemic of heart attacks and cancers.

When we hear something over and over, starting when we're young children, we accept it as true. For example, the myth that plant proteins are "incomplete" and need to be "complemented" for adequate protein is repeated over and over.[24] All vegetables and grains contain all eight of the essential amino acids (as well as the twelve other nonessential ones), although some vegetables have higher or lower proportions of certain amino acids than others. When eaten in an amount to satisfy our caloric needs, however, a sufficient amount of all essential amino acids are provided. Because digestive secretions and sloughed-off mucosal cells are constantly recycled and reabsorbed, the amino acid composition in postprandial (after-meal) blood is remarkably complete in spite of short-term irregularities in the dietary supply of amino acids.

In North America about 70 percent of dietary protein comes from animal foods. Worldwide, plants provide 84 percent of calories. It wasn't until the 1950s that human protein requirement studies were even conducted. These studies demonstrated that adults require 20 to 35 grams of protein per day.[25] Today, the average American consumes

100 to 120 grams of protein per day, mostly in the form of animal products—much more than necessary. People who eat a vegetable-based diet have been found to consume 60 to 80 grams of protein a day, still well above the minimum requirement.[26]

Advantages of Going Vegan or Very Close to Vegan

Even though eating an occasional small amount of animal product as flavoring or a condiment won't likely have a major effect on your diabetes control, there are other beneficial reasons to going all, or almost all, the way to a vegan diet. The main reason is that, for many people with diabetes, even a relatively low amount of animal protein in the diet could raise a hormone called insulin-like growth factor 1 (IGF-1). This is the main reason I am restricting intake to only six ounces per week.

IGF-1 is one of the body's important growth promoters in the womb and during childhood growth, but it also has anabolic (body-building) effects in adulthood. It is a hormone with a similar structure to insulin. The production of IGF-1 primarily takes place in the liver, and its production is stimulated by pituitary-derived growth hormone.

IGF-1 signaling is crucial for growth and development in childhood, but it promotes the aging process later in life. Reduced IGF-1 signaling is associated with enhanced life span.[27]

There is a tremendous amount of evidence regarding the life-span-enhancing effect of lower levels of IGF-1, especially in adulthood. Centenarians are known to be exceptionally insulin sensitive, which may protect against the insulin-resistance-associated, age-related increase in blood glucose levels. Lower levels of IGF-1 are associated with enhanced insulin sensitivity and enhanced life span.[28] This is critically important for people with diabetes or a tendency to develop diabetes, as higher levels of IGF-1 promote both diabetes and cardiovascular death from diabetes. The higher the biological value of the protein consumed, and the more of it consumed, the more IGF-1 produced. So the regular consumption of animal products is the most significant factor promoting IGF-1. Muscle tissue can produce its own IGF-1 in response to resistance exercise, but this does not raise systemic IGF-1 unless a diet rich in animal protein is consumed.[29]

IGF-1 and Cancer

The largest concern about elevated IGF-1 from our modern diet is its link to cancer. Elevated hormone levels caused by the Western diet are thought to contribute to the high rates of cancer in the modern world—not just sex hormones, such as estrogen and testosterone, but insulin and IGF-1 as well. The connection between increased IGF-1 signaling and cancer has been known for many years—in fact, cancer drugs targeting the IGF-1 pathway began to be developed in the late 1990s, and over seventy clinical trials have begun since then, many with encouraging results.[30] Because IGF-1 signaling plays a key role in tumor growth, reducing IGF-1 levels by dietary methods is now considered by most scientists studying this subject to be an effective cancer-prevention measure. IGF-1 signaling is involved in a number of processes relevant to tumor growth: proliferation, adhesion, migration, invasion, angiogenesis, and metastatic growth. A diet rich in antioxidants and phytochemicals results in reduced inflammation, oxidative stress, and IGF-1, which are critical to protecting against cancer and maximizing longevity.[31]

Protein Intake Promotes IGF-1

The composition of protein and the amount consumed also modify IGF-1 levels. Protein that is rich in the full array of essential amino

HIGH-CIRCULATING IGF-1 LEVELS HAVE BEEN LINKED TO SEVERAL CANCERS

Breast Cancer

The European Prospective Investigation into Cancer and Nutrition study found that elevated IGF-1 levels were associated with a 40 percent increased risk for women over the age of fifty.[32] In the Nurses' Health Study, high IGF-1 levels were associated with a doubled risk of breast cancer in premenopausal women.[33] Additional human studies, reviews of literature, and meta-analyses have also associated elevated IGF-1 levels with breast cancer.[34]

Colorectal Cancer

Elevated IGF-1 levels are associated with colorectal cancer, and IGF-1 promotes the spread of colorectal cancer cells.[35]

Prostate cancer

A 2009 meta-analysis of 42 studies concluded that elevated circulating IGF-1 is associated with increased risk of prostate cancer.[36]

Other Cancers[37]

Gynecological cancers
Multiple myeloma
Sarcomas
Renal carcinoma

acids causes larger increases in IGF-1 compared to protein not as biologically complete.[38] Plant sources of protein are less concentrated. They supply adequate protein, but not excessive amounts like animal products do, and the body needs to combine the amino acids for biological completeness, so they do not promote a surge in IGF-1 like animal proteins do. For example, milk and dairy products contribute to this excessive IGF-1 in circulation. In a meta-analysis of eight randomized controlled trials, circulating IGF-1 was found to be higher in milk-consuming groups compared to control groups.[39]

The Calorie Restriction Society is a collection of individuals who believe that consuming fewer calories will lead to a longer life. A six-year study of members of this group found that their IGF-1 levels were not significantly different from control groups on a standard Western diet (of course, body fat, fasting insulin, and inflammation markers were markedly lower in the calorie-restricted group). The Calorie Restriction Society group members were consuming an average of 108 grams of protein per day, far more protein than necessary. This led the researchers to then compare IGF-1 levels in members of the Calorie Restriction Society to vegans who had been consuming a moderately protein-restricted diet, averaging 50 grams of protein per day for at least five years. (One 3.5- to 4-ounce serving of chicken supplies 25–28 grams of protein.) The calorie intake was greater in the vegan group, but protein intake was lower and IGF-1 levels were indeed much lower.[40] This study cautions that overconsumption of protein, even when restricting calories, can keep IGF-1 levels elevated—to a point similar to those of typical Western eaters, who overconsume calories overall, blunting the potential of a longevity-inducing diet style. For example, many people eat egg whites believing that because they are almost pure protein and have no fat, they must be healthy. In truth, the high concentration of a biological protein makes egg whites disease promoting. Plant-based protein is much healthier.

Refined Carbohydrates Promote IGF-1

Although protein is the most important determinant of IGF-1 levels, excess intake of refined carbohydrates can also have an effect. Insulin regulates energy metabolism and affects IGF-1 signaling by increasing production of IGF-1 and decreasing IGF-1-binding proteins. It is likely that the Western diet increases IGF-1 via both excess protein and excess refined carbohydrate. Type 2 diabetes is associated with breast, colon, and pancreatic cancers, and there is evidence that insulin-mediated stimulation of IGF-1 production is partially responsible.[41] The take-home message here is to recognize that refined carbohydrates from processed foods and our nation's preoccupation with eating animal protein are both at the core of our cancer and diabetes epidemic. Until now, we have mistakenly focused on fat as the bad apple, endorsing egg whites and white meat, when actually these foods are not favorable for longevity. Note that switching to grass-fed beef or wild meats does not solve the problem with consumption of too many animal products, as the negative effects are not limited to fattened and sickly farm-raised animals. The heterocyclic amines, the heme iron, and the concentration of high biological protein are all negatives, especially for people prone to diabetes.

For many people, even a moderate amount of animal protein in the diet maintains unfavorably elevated IGF-1 levels and impedes the cholesterol-lowering and blood-sugar-lowering effects of a plant-based diet. But when we strive to consume most of our protein from plants, we solve the IGF-1 issue and help prevent both cancer and diabetes. The amino acids in plants are not as complete as those in animal products, so they do not raise IGF-1 to harmful levels, and they complement each other so we can achieve adequate levels of protein without going into excess.

Eating More Plant Protein Is the Key to Increasing Our Micronutrient Intake

It is interesting to note that foods such as peas, green vegetables, and beans have more protein per calorie than meat does. But what is not generally considered is that the foods that are rich in plant protein are usually the foods that are richest in nutrients and phytochemicals. By eating more of these high-nutrient, low-calorie foods, we get adequate protein and our bodies are simultaneously flooded with protective micronutrients. This fuels the reversal of diabetes and heart disease, helps heal the kidneys, and restores the body to a more youthful state. Animal protein is low-nutrient food. It does not contain antioxidants or phytochemicals, but plant protein does.

PROTEIN CONTENT FROM SELECTED PLANT FOODS	
FOOD	GRAMS OF PROTEIN
Almonds (3 ounces)	10
Banana	1.2
Broccoli (2 cups)	10
Brown Rice (1 cup)	5
Chick Peas (1 cup)	15
Corn (1 cup)	4.2
Lentils (1 cup)	18
Peas, frozen (1 cup)	9
Spinach, frozen (1 cup)	7
Tofu (4 ounces)	11
Whole Wheat Bread (2 slices)	5

When you drop body fat, your cholesterol lowers somewhat, but when you reduce or eliminate animal protein intake and increase vegetable protein intake, you lower cholesterol radically. This clearly is a vegetable-based diet, not one based in grains or animal products. Vegetables are rich in protein but also have almost no saturated fat or cholesterol, and they are higher in nutrients than any other food is. The cholesterol-lowering effect of vegetables and beans is without question. In addition, they contain an assortment of heart-disease-fighting nutrients independent of their ability to lower cholesterol. Amazingly, they also fight cancer. This food plan is designed to use large quantities of the most powerful anticancer, disease-fighting foods on the planet.

The point to keep in mind is that even if you completely ditch animal protein, or significantly minimize your consumption, you will still receive the protein your body needs through your vegetable-based diet.

The low-nutrient standard diet that is enjoyed by most Americans results in fatty deposits in the walls of the blood vessels. These eventually lead to blood vessel narrowing and blood clots that cause strokes and heart attacks. This occurs because of too many animal products, too many processed foods, and not enough natural, high-nutrient plant foods. The disease-building process is not the by-product of aging; it is the by-product of a diet poorly designed for humans. This diet gradually causes more and more damage as time goes on. Eventually, certain diseases and conditions crop up, mainly:

> Heart attacks and angina—diseased blood vessels in the
> heart (coronary arteries)
> High blood pressure and strokes—diseased blood vessels
> leading to and in the brain
> Dementia—diseased blood vessels in the brain
> Impotence—diseased blood vessels leading to and in
> the penis
> Claudication—diseased blood vessels in the legs

Unfortunately, the drug-favoring dietary advice typically offered to diabetics and heart patients is not science based, and it caters to Americans' social and food preferences and food addictions. In contrast, the nutritarian diet maximizes benefits for weight reduction, cardio protection, and diabetes reversal, effectively preventing and reducing the effects of all these conditions. And the food is delicious. With time, you will be shocked not only with the results but with the taste as well.

CHAPTER SIX

The Phenomenal Fiber in Beans

Susan Carno, an eighty-year-old female, was diagnosed with type 2 diabetes in 1987. She was on insulin therapy for twenty years, during which time she experienced a hypoglycemic episode (abnormally low glucose) at least once a month. Her blood sugar would be exceptionally high at times, and then in a few days, too low. She gained a significant amount of extra weight while she was on insulin. She also had a history of high cholesterol and high blood pressure and a family history of heart disease, diabetes, and hypertension. During the last year before becoming my patient, she was experiencing increased frequency of hypoglycemic reactions. A grand mal seizure attributed to the hypoglycemia finally convinced her that she needed to do something different and motivated her to find me.

Susan's physician had initially prescribed statin drugs for high cholesterol, but when Susan developed muscle aches, the medication was discontinued. Besides taking insulin (30 units of Lantus once a day and 5 units of Humalog with each meal), she was also taking metformin and Byetta injections for diabetes.

*I talked to Susan on the phone briefly almost every day to
check her morning glucose number, and she was able to cut
back on the insulin more quickly than I expected. Originally
I thought that she would still need some insulin even on this
diet because she was eighty years old and her pancreatic beta
cell reserve was likely compromised. I was wrong. Once she
started my nutritarian approach for diabetes, I was able to
discontinue all of her insulin within the first ten days. The
diet was more powerful than 45 units of insulin, even with
no change in her activity level.*

*Susan was able to discontinue all of her diabetes medica-
tions by the end of the first month. Her HbA1C went from
7.3 to 6.6 with no drugs. It takes three months for HbA1C
to reflect new low glucose readings, and at the three-month
checkup, on no medications, her glucose was running around
100. She also had no further hypoglycemic reactions. She lost
thirty-eight pounds, going from 148 pounds to 110 pounds.
Her blood pressure declined from 172/82 to 130/75. Her
total cholesterol did not change much, but her cholesterol/
HDL ratio improved from 4.0 to 3.3, and her triglycerides
improved considerably.*

*Susan was thrilled. While taking insulin, she'd felt like a
prisoner inside her own body, aging rapidly. It made her feel
hungry and overall left her feeling constantly sick. For twenty
years, she injected herself because she saw no other option.
When the insulin stopped, Susan suddenly discovered she
had a great deal more energy. Over the year after we started
working together, Susan's exercise endurance increased
considerably and she was able to comfortably walk for a
full hour or more. She has been enjoying sixty-minute walks
every day. After being on the diet for a year, she celebrated
with her favorite ice cream. Surprisingly, she did not like it*

as much as the fresh fruit sorbet I taught her how to make.
It was just too sugary for her, so after the first few spoonfuls
she stopped!

Beans, green vegetables, seeds, and some fruits are high in soluble fiber. Soluble fiber supplies a gelatinous-like material in the bowel. It is not absorbed and does not give us calories. Soluble fiber is very important, as it slows the absorption of glucose and helps lower cholesterol. Beans are especially high in soluble fiber.

Insoluble fiber—roughage—is important too. It provides bulk to our stool and keeps us regular. And guess what: seeds, nuts, vegetables, and, yes, beans have plenty of insoluble fiber too.

For years, nutritionists and scientists thought there were only two kinds of fiber—soluble and insoluble. Now we know there is a carbohydrate that acts like a fiber too. It is called resistant starch. It supplies few calories, and most of the calories do not raise glucose levels. It is called resistant starch because it is resistant to stomach acid and digestive enzymes. It is not digested in the small intestine but passes to the large intestine, where it undergoes fermentation. Fermentation means that the bacteria decompose and degrade this starch into simpler compounds. When the bacteria in the bowel degrade the resistant starch, it forms new compounds that have health benefits. Resistant starch is important for good health and has beneficial effects for diabetics.

Legumes such as beans, lentils, peas, and chickpeas fall far below grains on the list of foods Americans eat. However, legumes are richer in nutrients, protein, and fiber, and they contain much higher levels of resistant starch. Considering their favorable effects on blood sugar and weight loss, they are the preferred carbohydrate source for people who have diabetes or are at risk for diabetes.

Most starchy foods have a small amount of resistant starch in them. At the beginning of human history, fruits contained more resistant starch and fiber and less sugar than fruits that are commonly available now. Wild foods, the same as those early human foods, are more fibrous compared to what is bred, cultivated, and processed today. If you were to taste a wild pineapple, wild lychee, or wild plantain in the tropical jungle, you would find that any of them are hardly sweet, much chewier, and fibrous—and filling from all the fiber—but they are certainly not calorically dense. It is certain that in a primitive tropical habitat that provided a diet of just wild food and greens and maybe some fish, there would be no overweight or diabetic people. If you were ever shipwrecked on a deserted island, it would be almost impossible to become or remain overweight.

Proponents of high-protein, low-carbohydrate diets argue that intake of carbohydrates—especially starch—should be restricted or eliminated and substituted with animal products instead. This might seem logical on a superficial level, but when you look deeper into the science, you find that fiber-less animal products contribute to diabetes-related health risks, while beans—even though they are largely carbohydrates—directly lessen these risks and promote the reversal of diabetes.

Dietary starch is most often converted to glucose. When not burned as energy for immediate use, it is stored as glycogen, a high molecular-weight polymer of glucose. The body is capable of storing approximately 300 to 500 grams of glycogen at one time. Any excess glucose that is not rapidly burned as fuel or stored as glycogen is converted to fat and stored as body fat. Meat-based-diet proponents argue that to lose weight, we should eat less starch. They are right to a degree. Certainly we should eat less high-glycemic, low-nutrient starch, and certainly we should not overeat. When we eat mostly high-starch foods, especially refined carbohydrates, it promotes swings in blood glucose, putting excess work on the pancreas to produce a huge insulin load. Plus if we overeat, our glycogen stores could be already full, meaning the extra carbohydrate calories we don't need will be stored as fat on the body. But not all carbohydrates fall into this high-glycemic, low-nutrient starch category.

The Whiter the Bread,
the Sooner You're Dead

Some people think that sugar-free cookies, cakes, and pastries can actually help their diabetes or help them lose weight. This is not the case—these sugar-free products are essentially low-nutrient junk foods. White flour actually makes your blood sugar rise almost as much as plain sugar does. Carbohydrates are chains of sugar molecules lined in a row. They are found in all plants and foods made from plants. Carbohydrates can be a single sugar, or three or four bound together, but when thousands of sugars are bound together, they are called starch. When these simple carbon molecules are bound together so tightly that your body cannot break them down and digest them, they are called fiber.

Only simple sugars can pass from your intestines into your bloodstream. When your digestive enzymes break down the carbohydrates into simple glucose molecules, they are absorbed immediately and enter the body just as if you had sucked on a sugar cube. Indeed, eating sugar and white flour does not cause just diabetes; these foods are also linked to heightened risk of cancer. Quite a few studies have linked the consumption of high-glycemic, low-nutrient food to cancer. One study showed over a 200 percent increase in risk of breast cancer in women eating more than half their diets as refined carbohydrates.[1] Too many Americans still eat this type of diet, with more than half their calories coming from processed foods. These individuals are slowly destroying their health. Eating processed foods is like snorting cocaine. Eventually you will pay a big price—your health. And the more a person consumes this deadly white stuff, the stronger the cravings for more.

Bagels, white bread, pasta, pizza, and rolls are all staples of the American diet, and they are a large contributor to our epidemic of obesity, diabetes, heart disease, and cancer. Commercial wheat products are also treated with fungicide, sprayed with insecticides, and bleached with chlorine gas or other chemicals. They return little nutrient bang for all their calories. To put it bluntly, these staples of our diets are disease-promoting junk food. All the white starches—basically, white bread, white rice, and even white potatoes—are very rapidly converted to glucose, which is sugar, and absorbed into the bloodstream, shooting blood sugar levels up.

When blood sugar skyrockets, it overworks the pancreas to match the load of sugar with a large amount of insulin. This is not only stressful to the body and the pancreas, but metabolizing that large energy load without a concomitant intake of micronutrients creates metabolic havoc in the cells. Toxic metabolites build up in cells when we consume calories without antioxidant and phytochemical micronutrients needed to remove and control the toxic by-products. So as we eat more low-nutrient and low-fiber carbohydrates, we build up more cell toxicity, leading to disease and food addiction.

Most of the common carbohydrates we eat are turned into glucose, but it is important to realize that the conversion efficiency and rate vary greatly from one type of carbohydrate to another. For example, the starch in potatoes, cereals, and baked goods digests very rapidly; all their calories are converted quickly, supplying the body with a huge glucose load. The starch in beans, barley, and black wild rice is digested more slowly and causes a much slower and lower blood sugar rise. Beans are at the top of the preferred carbohydrate totem pole because they contain more of both slowly digestible starch and resistant starch.

Unique properties of the carbohydrates in beans and legumes include:

- Higher amount of slowly digestible starch

- Higher amount of resistant starch

- Higher amount of insoluble fiber

- Higher amount of soluble fiber

Resistant starch actually goes all the way through the small intestine without being digested at all. In this way, it is more like fiber, and in some cases is classified as a type of insoluble fiber.

Resistant Starch Is the Secret

There are different types of resistant starch in foods. Amylose and amylopectin are examples. It is starch that is tightly packed in a stable crystalline form within foods, making it difficult to digest. The more resistant starch that reaches the colon undigested, the less calories we absorb from that food. When resistant starch reaches the colon, the bacteria there use it for fuel. The resistant starch is also, therefore, a prebiotic, meaning it serves to fuel the growth of beneficial bacteria in the colon.

This process of degrading these starches by bacterial action is called fermentation, and it produces a type of fat called short-chain fatty acids (SCFAs). In other words, the resistant starch does not even get converted to a simple sugar; it gets converted into a simple fat. Only a small percent of these calories are absorbed by the body, but they are highly beneficial.[2] So calories from resistant starch are listed on the food labels but almost 90 percent of those calories do not get absorbed and they do not raise blood sugar at all. Resistant starch is especially associated with one type of SCFA called butyrate.

Now here's the fascinating part: even though only a small amount gets absorbed, butyrate offers a wide array of health benefits, including strong protection against colon cancer. It protects our bodies in lots of other ways too, namely by enhancing the absorption of beneficial minerals like calcium and magnesium and so, importantly, improving insulin sensitivity. It has the opposite effect of eating sugary or high-glycemic starches. It actually improves diabetic glucose numbers the day after it's eaten.[3]

Most importantly, these SCFAs slow down glycolysis in the liver, thus delaying hunger, and they increase the breakdown of body fat as a source of energy, facilitating weight loss. As you recall, glycolysis is the breakdown of the stored glycogen back into glucose for use by the body. The small amount of SCFAs that are absorbed increases fat oxidation, meaning your body burns fat for energy more efficiently, encouraging weight loss.[4]

When you eat a meal of mostly green vegetables, eggplant, onions, mushrooms, and a cup of beans, biochemical events occur that work medicinally; they repair the biochemical defects that would lead to diabetes. In fact, in direct contrast to meat and potatoes, just one additional serving of green vegetables in a diet has been demonstrated in a meta-analysis to offer significant diabetes protection independent of the effects on weight reduction.[5] The authors of the study speculate these profound benefits were due to the high levels of beneficial micronutrients in greens. Then you add beans, and more magic happens.

> Beans store well and are inexpensive,
> highly nutritious, and entertaining.

Let's review some of the benefits of eating greens and beans, instead of bread, rice, and potatoes:

- Increased nutrients, SCFAs, bacterial activity, and fiber, which lower cholesterol and triglycerides[6]

- A sensation of fullness or satiety, leading to meal satisfaction even though fewer calories are consumed

- Lower glucose level and improved insulin sensitivity, aiding in the reversal of diabetes

- Promotion of good bacteria, which aid nutrient absorption; suppression of bad bacteria and their toxic products

- Bowel regularity, preventing constipation

- Efficient removal of cholesterol and other unhealthful fats
- Less fat stored after meals
- Increased fat breakdown
- Slowed glycogen utilization in the liver, which delays hunger and promotes a lower calorie intake
- Protection against colon cancer development

Beans are the very best source of resistant fiber. Although the types of beans and preparation methods cause varying amounts of resistant starch (canned beans are more glycemic), in general, the starch in beans is about evenly divided between slowly digested starch and resistant starch. Note, though, that taking products such as Beano to increase digestibility of beans will also increase the digestibility of resistant starch and enhance its caloric absorption. Instead, it is better to allow the body to adjust to the use of beans in the diet over time, then, gradually, favorable bacteria will increase in number in the digestive tract, which will facilitate the digestion of the resistant starch in the beans.

Black beans, for instance, contain the highest amount of total dietary fiber at 43 percent, and 63 percent of their total starch content is resistant starch. Cereal grains, especially barley and corn, follow legumes in their percentages of resistant starch that reach the colon, but they drop off significantly in fiber content. Heavily processed flours and grain-based products have a very low resistant starch content with a range of 1.7 percent in white flour to about 4.6 percent in pasta. The point here is that the major source of carbohydrates in the diabetic diet should be beans, not grains or root vegetables like potatoes. I sometimes call my diabetic diet recommendations the greens and beans diet.

The resistant starch found in beans powerfully reduces hunger and, thus, food consumption over many hours, coinciding with the fermentation that takes place in the large intestine hours after eating the beans. So eating beans with lunch will reduce your hunger and appetite for dinner many hours later, overall lowering the amount of calories you desire for the day. For diabetics, beans are critical for lowering the insulin requirement for starch digestion. They also supply amino acids that complement the other vegetables, nuts, and seeds to enhance the biologic value of the protein in the diet, without raising IGF-1.

Refer to the food data chart. If you add up the percent of resistant starch (RS) (90 percent of which is unabsorbed) and fiber (which also is not absorbed) and make comparisons with other starches, you will see why beans are the preferred starch for diabetics. The RS + fiber is a handy measurement to illustrate the favorable effects of beans for diabetics compared to other high-carbohydrate foods, and certainly it is a much more critical measurement than the glycemic index for diabetes management and weight loss.

Let's not forget that the ANDI scores of beans are high as well. They have been found to contain significant levels of polyphenols, which have anticancer effects. Red and black beans have been found to produce apoptosis (cell death) of colon cancer cells. This means the compounds in beans and those produced in the digestion of beans have beneficial effects to fortify cells against cancer. They also cause cells that have become precancerous or cancerous to die off before they can actually multiply into cancerous tumors. They protect people against colon cancer—the exact opposite of what red meat does. The Polyp Prevention Trial demonstrated that beans provide more protection against advanced adenomas of the colon than any other food does, with a 65 percent reduction of adenomas in participants in the highest quartile of dried bean intake.[7] And people who eat beans merely twice a week were found to have about a 50 percent reduction of colon cancer.[8] Imagine the protection we would achieve if we ate beans almost every day in conjunction with other well-investigated cancer-fighting foods.

Other sources tout beans' life-lengthening benefits too. The conclusions of an important longitudinal study show that a higher legume intake is the most protective dietary predictor of survival amongst the

FOOD DATA*	RS %	FIBER %	RS% + FIBER%
Black Beans	26.9	42.6	69.5
Northern Beans	28.0	41.1	69.1
Navy Beans	25.9	36.2	62.1
Red Kidney Beans	24.6	36.8	61.4
Lentils	25.4	33.1	58.5
Split Peas	24.5	33.1	57.6
Black-eyed Peas	17.7	32.6	50.3
Corn	25.2	19.6	44.7
Barley	18.2	17.0	35.2
Brown Rice	14.8	5.1	20.5
Millet	12.6	5.4	18.0
Rolled Oats	7.2	10.0	17.2
White Rice	14.1	1.5	15.6
Whole Wheat Flour	1.7	12.1	13.8
Pasta	3.3	5.6	8.9
Potato Flour	1.7	2.1	3.8

* The percentage of starch that is resistant starch varies considerably dependent on preparation method or form of the food studies, and measurement methods. It is clear that beans/legumes are the richest sources of resistant starch in the diet, regardless of cooking method. The study that supplied this data utilized a consistent foods preparation method so reliable comparisons can be made.

elderly, regardless of their ethnicity, in multiple populations studied.[9] The study found that legumes were associated with long-lived people in various food cultures including Japanese (who eat beans in the form of soy, tofu, natto, and miso), Swedish (brown beans and peas), and Mediterranean (lentils, chickpeas, and white beans) diets. Beans and greens are the foods closely linked in the scientific literature with protection against cancer, diabetes, heart disease, stroke, and dementia.

We are not done with beans yet. Because beans contain so much resistant starch, as much as 20 grams in each cup of navy beans (4 calories per gram), and because they contain amylase inhibitors that resist the

Edamame, A Super Bean

Edamame beans are young green soybeans that are boiled in the pod and served hot or cold. They are usually served in the pods, but you squeeze them out of the inedible pods to eat them. They are available in supermarkets, mostly in the frozen vegetable section, both shelled and in the pods. Young green soybeans are an unprocessed, natural food that is rich in minerals, calcium, and omega-3 fats and extremely high in protein.

Frozen pods can be boiled for five minutes or just defrosted overnight in the refrigerator. They are a unique bean because they are so low in carbohydrates. They are a great food for diabetics and a delicious addition to a salad or vegetable dish. Multiple scientific studies have demonstrated that unprocessed soy beans have dramatic health benefits. Besides lowering cholesterol and blood glucose, they also prevent breast cancer.[10]

digestion of their starch, a decent percentage of the calories that are listed on the food label do not actually come into the bloodstream as glucose or even as calories. This is complicated, but remember that when the resistant starch hits the colon, it is acted on by bacteria and transformed into SCFAs. That transformation results in only about 2 calories per gram. The fermentation occurs so low down in the digestive track that very little of it gets absorbed. This resistant starch, as well as most of the other fibers that are not assimilated, are counted as a part of the 200 to 300 calories per cup in the nutritional info. But the bottom line is that even though a cup of cooked beans may be listed as 225 calories, they actually give you much fewer calories per cup, a higher percent of protein, and fewer carbohydrates than show up in their analysis. All those listed 225 calories are setting off caloric and nutrient receptors in the stomach and small intestines, registering satiation and telling you that you have eaten enough. And here's the amazing thing, because of the presence of the amylase inhibitors and the resistant starch, maybe a quarter of those carbohydrate calories don't even get absorbed.[11]

So when you make beans your favored carbohydrate source, you:

Get high-quality protein

Get a powerful anticancer food

Get more satiation that turns off your desire to eat
more food

Burn more fat, helping you get rid of your diabetes

Have better digestion

A healthy pantry needs to contain a stock of dried beans and canned beans. If your local market doesn't have unsalted canned beans, you can usually find them in a health food store. If you have no other option but salted canned beans, rinse them before using to remove about half the excess salt. Dried beans that are first soaked overnight and then cooked are the best way to use beans. They don't have to be soaked in advance, but it speeds up the cooking time.

Dried beans are the most economical high-nutrient food. You save money on food when you use lots of dried beans in your cooking. Also, sprouted whole beans and grains offer great nutrition and save money on your food bills. Soak beans overnight in a jar, and then rinse and drain the water out every day for the next four to six days, and you will have bean sprouts to use in your salad or in a vegetable dish.

Split peas, squash (butternut, acorn, or winter), corn, wild rice, quinoa, and wheat berries are healthy high-carbohydrate foods that contain a moderate to intermediate amount of resistant starch but are still fairly nutrient-rich foods. Dark or black wild rice is more fibrous and higher in resistant starch compared to ordinary brown rice. Note that wheat berries and coarsely ground sprouted wheat are more favorable than whole wheat flour, especially whole wheat pastry flour. The more finely ground the grain, the higher its glycemic load, and the more it is cooked, the more diabetic unfavorable it becomes.

Avoid anything that is white—white flour, white pasta, white potatoes, and white rice. Other carbohydrate-rich foods can be used in moderate quantities depending on body weight and diabetic parameters.

When looking at a choice of various high-carbohydrate plant foods, as a diabetic, consider:

1. The fiber content

2. The percent of slowly digestible starch

3. The percent of resistant starch

4. The micronutrient content

5. The caloric density

6. The beneficial qualities of other foods in the menu that may have to be eliminated or reduced to allow room for this food

There is a nutritional hierarchy of carbohydrate-rich plant foods. Beans, cauliflower, and other more nutritious high-carbohydrate foods are most heavily emphasized because of their micronutrient density, fiber, low GL, slowly digestible starch, and resistant starch content. Interestingly and conveniently, the micronutrient density of high carbohydrate plant foods parallels their fiber and resistant starch content.

The GL is also an important component to consider in creating the best diet for those with diabetes or prone to it. As we discussed, diets containing large quantities of high GI foods are associated with the risk of diabetes, heart disease, multiple cancers, and overall chronic disease.[12] GL may not be the main thing, but it should not be completely ignored. It is likely the main reason why white potatoes have been linked to a worsening of glycemic control in diabetics.[13] In fact, when 84,555 women aged thirty-four to fifty-nine were tracked for twenty years in the Nurses' Health Study, researchers found that regular potato and French fry consumption was significantly associated with increased risk of diabetes; substituting a whole grain for a serving of potato per day lowered risk of diabetes by almost 30 percent.[14]

Examining the glycemic effects of various carbohydrate choices enables us to modify food choices for improved glycemic control and enhanced weight loss.[15]

GLYCEMIC LOAD OF COMMON HIGH-CARBOHYDRATE FOODS[16] (1 CUP)	
Cauliflower	negligible
Split Peas	4
Black Beans	6
Red Kidney Beans	7
Butternut Squash	8
Green Peas	8
Beets	9
Lentils	9
Whole Wheat	11
Barley	13
Navy Beans	13
Rolled/Steel-cut Oats	13
Black Eyed Peas	14
Quinoa	16
Corn	18
White Bread (2 slices)	20
White Pasta	21
Brown Rice	24
Millet	26
White Potato	29
White Rice	29
Cola (16 ounces)	32

It is reasonable to limit white flour, white potato, and rice intake considering the large variety of carbohydrate-rich plant foods (beans, intact whole grains, cauliflower, peas, squash, and other starchy vegetables) to choose from that are more nutrient dense and without such a high GL. In general, I recommend diabetics avoid regular and liberal consumption of foods with a GL above 15, at least until their diabetes is in better control and their weight has dropped significantly. Remember, *intact whole grains* refers to most grains that are not ground into flour. If you want to eat a food made from flour, like pasta, use an alternative like black bean pasta or lentil pasta. You will be amazed by its taste.

My results with thousands of clients and patients indicate that excessive consumption of white potatoes (even prepared healthfully) and white rice may be problematic due to their high GL, especially in overweight individuals and those with diabetes and prediabetes. With some consideration of the factors discussed here, diabetics can remedy their challenges such as residual body fat, high triglycerides, suboptimal lipid profiles and nonoptimal fasting glucose levels. They also glean other benefits, especially heightened protection against cancer due to the inclusion of more cancer-protective micronutrients in the nutritarian diet.

So for maximizing weight loss and reversing diabetes, the trick is to use beans as your primary starch source. Eat some beans in a salad or

THE BEST FOODS TO EAT TO REVERSE DIABETES

Raw greens	Tomatoes
Cooked greens	Cauliflower
Mushrooms	Beans
Eggplant	Nuts, Seeds
Onions	Lower-sugar fruits such as berries and kiwi

soup with lunch and again with a green vegetable for dinner. Eat other non-bean, starchy foods such as squash, green or split peas, water-cooked steel-cut oats, black (wild) rice, or wheat berries in small amounts with breakfast or as part of a dinner vegetable casserole, but do not eat them as the main volume of food with the meal. Remember, incorporate these whole grains only in a limited amount in the diet until weight loss and glucose readings are under control.

What About the Gas?

The body can learn to adjust the bacteria in the digestive tract to better digest beans and avoid gassiness. It just takes some time. Start out with a tablespoon of beans, chewed well, at each meal. In other words, have beans all the time, but begin with a nonthreatening amount so you do not become a gas factory. In a few weeks, you will be able to gradually increase the amount without experiencing digestive difficulty. Lentils, split peas, and garbanzo beans are less gassy because they have less resistant starch. So start with those first and gradually move up to red and black beans that are even more favorable for diabetics. Do not take digestive enzymes or Beano; instead take a probiotic such as lactobacillus acidophilus. Don't forget to chew the beans very well, as most gas production comes from poorly chewed food.

Most people on my antidiabetic diet enjoy a salad and a bean-based vegetable soup for lunch. The beans used are the dry beans that have been cooked in the soup base. For dinner you can toss some beans on top of your salad or mix them into a cooked green vegetable dish. Beans offer a wide variety of flavors and textures and also make great dips, sauces, and chewy bean burgers.

CHAPTER SEVEN

The Truth About Fat

*My blood sugar was always out of control until I started
eating the way Dr. Fuhrman showed me. My endocrinologist said there had to be a mistake when he saw my HbAIC
below 6.0. I feel much better eating this way, because when
blood sugar is up and down, it makes you feel bad.*

—Robert, age 11

*When I got diabetes, my mom took me to Dr. Fuhrman, and
now my whole family is healthier. I thought I couldn't live
without bagels and pizza, but I learned to like the foods that
are good for me. When my mom took away the food I was
used to, I thought I was going to starve to death, but now I'm
happy she did it.*

—Janice, age 13

The major determinant of our long-term health is the nutritional
quality of the calories we eat. It is the quality of the fat, the quality
of the protein, and the quality of the carbohydrates we eat that most
influence our health.

Ask yourself: *Is the food I am about to eat a whole, natural plant source of calories? Is it packed with fiber, antioxidants, and phytochemicals? Does it contain not only known nutrients but plenty of undiscovered nutrients too? Or were most of those fragile, but beneficial, nutrients lost in the way the food was processed or prepared?* These are even more important questions than whether it is a low-fat or high-fat food.

You may have heard that nuts and seeds are fatty and fattening and are foods to be shunned. However this myth is finally buried.[1] Recent evidence from several studies shows a wide variety of health benefits from eating these foods. There has never been a study that showed any negative health outcomes from consuming these natural, high-fat, whole plant foods. In fact, the studies show only positive health benefits and conclude that these foods should be an important part of a well-rounded, healthy diet. It must be emphasized that health problems associated with high-fat diets are from consuming animal fats, processed oils, and trans fats, not from avocados, raw nuts, or seeds.

As discussed, macronutrients—fat, carbohydrates, and protein—are the three sources of calories. Americans eat too much of all three. I intentionally do not give a preferred percentage of each macronutrient in the diet, and I do not recommend that fat be (exclusively) limited. Trying to micromanage the precise amount of each caloric source misses the most critical issue in human nutrition. The real critical issue in human nutrition is meeting your macronutrient needs without excess, while getting sufficient micronutrients (vitamins, minerals, and phytochemicals—the parts of food that do not contain calories) in the process. There is a broad acceptable range in the macronutrient ratio as long as we are not overeating calories. Certainly if we're overweight, we want to be more careful and limit consumption of these higher-fat foods. As we have seen, it is easy to overeat high-fat foods, as they are a concentrated source of calories. The goal is to find the right balance. Adhering to a diet that is less than 10 percent of calories from fat is not an appropriate recommendation for ideal health. This too-low percentage of fat results in less-than-ideal health outcomes such as low energy and minimized hormone production. The simple truth is that a healthy diet could be 15 percent of calories from fat or even 30 percent of calories from fat. As long as the diet is rich in micronutrients and does not exceed our daily need for calories, a lower-fat diet has no advantage in the prevention and treatment of disease.

> There is *no* evidence to suggest that a diet of equal calories that is extremely low in fat is an advantage for prevention or treatment of heart disease or any other disease. Studies that compare dietary fat percentages suggest that it is not the fat level, but other more critical qualities, that make the diet more or less beneficial.

I want to be clear that the benefits of a vegetarian or vegetable-based diet are not the result of low-fat intake. Most vegan or vegetarian diets are not ideal because they lack green vegetables. It may seem odd that I am claiming most vegetarian diets are lacking in high-nutrient vegetation, seeds, and nuts—all healthful vegetarian foods—but all too often, that is the case. Achieving an ideal level of phytonutrients and other micronutrients necessitates eating a large amount of green vegetables each day. Any diet that does not contain sufficient vegetables is lacking. When you eat lots of vegetables, especially green vegetables, you meet your body's need for fiber and micronutrients with very few calories. Then to balance the diet and fill your caloric needs, you can choose an assortment of other foods, preferably ones that are of high nutrient quality. Unlike other people advocating plant-based diets, I recommend more vegetables, fruits, beans, nuts, and seeds and less bread, potato, and rice. The daily addition of one or two ounces of nuts and seeds, which average about 175 calories an ounce, can bring the diet up to 15 to 30 percent of calories from fat. This is important, and

I repeat: my recommended diet is eating 15 percent or more calories from fat in the form of healthy, whole foods, not oil.

It might seem logical to restrict higher-fat foods like nuts, seeds, and avocadoes because high-fat foods are higher in calories (fat is 9 calories per gram compared to 4 calories per gram in carbohydrates and protein). Although you should take care not to eat too many calories and to adjust the level of these foods to maintain a slim body, there are lots of good reasons to include at least some of these higher-fat foods in your diet. Accumulating research shows that a diet as low as 10 percent of calories from fat may be too low, even for the overweight, diabetic, or heart disease patient. Judicious use of these higher-fat foods is beneficial for not just heart disease, but for weight loss and diabetes too.

The scientific literature corroborates my clinical experience over the last twenty years as I have cared for thousands of patients with obesity, diabetes, and heart disease. It provides evidence to show that for every calorie from white rice, potato, white bread, or animal products that is removed from the diet and substituted with raw seeds and nuts, there are many health benefits, including:

- Lower blood sugar
- Lower cholesterol
- Lower triglycerides
- Better LDL/HDL ratio
- Better antioxidant status
- Better absorption of phytochemicals from vegetables
- Better diabetic control
- Lower weight
- More effective reversal of heart disease
- Prevention of cardiac arrhythmias in heart patients

- More weight loss, not weight gain
- Better nutritional diversity and satisfaction with fewer calories
- Increased protection against cancer
- Better muscle and bone mass with aging

Nuts and Seeds Protect Against Cardiovascular Death

Raw nuts and seeds are packed with nutrients. They contain lignans, bioflavonoids, minerals, and other antioxidants that protect the fragile freshness of the fats within and contain plant sterols and proteins that naturally lower cholesterol. Because these foods supply certain fibers, phytochemicals, phytosterols, and bioactive nutrients like polyphenols and arginine that are not found in other foods, they also prevent blood vessel inflammation.

Perhaps one of the most unexpected findings in nutritional epidemiology in the past five years has been that nut consumption offers strong protection against heart disease. Several clinical studies have observed beneficial effects on blood lipids as a result of diets high in nuts (including walnuts, peanuts, almonds, and other nuts).[2] A review of twenty-three intervention trials using nuts and seeds convincingly demonstrated that eating nuts daily decreases total cholesterol and LDL cholesterol.[3] Not only do nuts and seeds lower LDL (bad) cholesterol, but they also raise HDL (good) cholesterol. Interestingly, they can help normalize a dangerous type of LDL molecule—the small, dense LDL particles that are particularly damaging to blood vessels, especially the endothelial cells lining the blood vessels.[4]

Ellagitannins are dietary polyphenols with potent antioxidant and other cancer chemopreventive activities. They are found in berries, nuts, and seeds and are best absorbed from walnuts.[5] Walnuts can reduce

C-reactive protein and harmful plaque adhesion molecules, two significant markers of inflammation in arteries. The result is improved, and even restored, endothelial function (which includes the elastic property of arteries to dilate when meeting an increased demand for blood). According to the researchers, walnuts are the first food to show such cardiovascular benefits.[6]

Studies on nuts reveal much more than just their power to change risk factors like blood glucose or cholesterol; they actually show that nuts decrease cardiovascular death and overall increase life span.[7] Five large prospective cohort studies (Adventist Health Study, Iowa Women Health Study, Nurses' Health Study, Physicians' Health Study, and CARE Study) have studied the relationship between nut consumption and the risk of atherosclerotic heart disease.[8] All found a strong inverse association. That means nuts are lifesaving.

The protective effect of nut consumption on heart disease is not offset by increased mortality from other causes. In fact, nut consumption has been found to be inversely related to all-cause mortality in all tested populations including whites, blacks, and the elderly. Eating nuts and seeds offers a well-documented intervention for increasing longevity. The beneficial effects of nut consumption observed in clinical and epidemiologic studies underscores the importance of distinguishing different types of fat. Nuts contain mostly mono- and polyunsaturated fats that lower LDL cholesterol levels. However, the favorable fat issue does not alone account for the health benefits of nuts and seeds. These powerful health benefits are not achieved when oils, rather than whole nuts and seeds, are substituted as a caloric source.

Based on data from the Nurses' Health Study, it was estimated that substituting fat from one ounce of nuts for equivalent energy from carbohydrates in an average diet was associated with a 30 percent reduction in heart disease risk. The substitution of nut fat for saturated fat was associated with a 45 percent reduction in risk. Frank Hu, M.D., a leading researcher at the Harvard School of Public Health, on the value of nuts in the American diet, says, "Our epidemiological studies have shown eating about one ounce of nuts a day will reduce the risk of heart disease by over 30 percent."[9] The Physicians' Health Study added much more to the story. The most fascinating and perhaps most important finding is that nuts and seeds do not just lower cholesterol and protect against heart attacks. Components of nuts apparently also have antiarrhythmic and antiseizure effects, which dramatically reduce the occurrence of sudden death.[10] These rhythm-stabilizing effects of nuts and seeds are due not only to the amount of omega-3 fatty acids but also to other beneficial qualities of these natural foods.

The Physicians' Health Study followed 21,454 male participants for an average of seventeen years. Researchers found a lower risk of sudden cardiac death and other coronary heart disease end points after controlling for known cardiac risk factors and other dietary habits. When compared with men who rarely or never consumed seeds or nuts, those who consumed two or more servings per week reduced their risk of sudden cardiac death by over 50 percent. This means that the consumption of nuts powerfully reduces the chance of having a life-threatening cardiac arrhythmia called ventricular fibrillation or ventricular tachycardia. People who have heart disease do not always die of heart attacks; they often die of an irregular heartbeat that prevents the heart from pumping properly.

The absence of nuts and seeds in a diet may actually increase the risk of one of these fatal heart rhythm disturbances. In my years of medical practice, the most common reasons patients have come to see me are high blood pressure, high cholesterol, atherosclerosis, angina, diabetes, and being overweight. People following my nutritional advice have seen dramatic improvements in their conditions. They have lost weight, normalized their blood pressure and cholesterol, and reversed their heart disease and atherosclerosis in an impressive and often dramatic fashion. All my patients were advised to eat some raw nuts and seeds in their diets.

For the Reversal of Diabetes and Obesity

Epidemiologic studies indicate an inverse association between frequency of nut consumption and body mass index. Interestingly, their consumption may actually suppress appetite and help people get rid of diabetes and lose weight.[11] In other words, people who consume more nuts and seeds are likely to be slim, while people consuming fewer seeds and nuts are more likely to be heavier. Well-controlled nut-feeding trials, designed to see if eating nuts and seeds resulted in weight gain, showed that eating raw nuts and seeds promoted weight loss, not weight gain. Several studies have also shown that eating a small amount of nuts or seeds actually helps dieters feel satiated, stay with the program, and have more success at long-term weight loss.[12]

Oil is fattening, containing 120 calories per tablespoon, and can sabotage your opportunity to lose weight or reverse your diabetes. Plus, it does not have the protective effects on your heart or diabetes. This program uses seeds and nuts, not oil, as the main fat source, and to flavor dressings and dips.

Because nuts and seeds are rich in minerals and fiber and have a low GI, they are favorable foods to include in a diet designed for diabetics and those with prediabetic symptoms. Researchers from Harvard noted that people eating one ounce of nuts five times a week reduced their risk of developing diabetes by 27 percent.[13] The features of a diet that make it favorable to diabetics are not adequately described by the terms *vegan* or *low fat*. There are many vegan foods and vegan diets that would be unfavorable for diabetics, especially those that include lots of oil, finely ground grains, and foods made from white flour and white potatoes. The qualities of a diet that make it maximally favorable to diabetics are:

Overall calories and weight loss results

Amount of fiber consumed per meal

Micronutrient diversity and completeness

Glycemic load of the meals

Antioxidant and phytochemical index

Satiety and removal of food cravings and addictions

A vegan diabetic study "A Low-Fat Vegan Diet Improves Glycemic Control and Cardiovascular Risk Factors in a Randomized Clinical Trial in Individuals with Type 2 Diabetes" was published in *Diabetes Care*.[14] The word *vegan* did not adequately describe the features of the diet that made it more favorable compared to the ADA diet. The researchers were careful to remove all vegetable oils and white flour products. If they'd called the diet a high-fiber, no-processed food, vegan diet, it would have been a more descriptive title indicating why it was somewhat more effective compared to the ADA diet. However, the results still did not compare in effectiveness with the program I am teaching here.

Typical vegan diets do not show the dramatic improvements in lipids, triglycerides, glucose, and even weight loss. One important design feature for better health and disease reversal is the reduction of high-starch vegetables and grains and the substitution of beans, nuts, and seeds instead. For example, another representative study showed that women on a low-fat vegan diet lowered LDL cholesterol by 16.9 percent.[15] In a similarly conducted study including nuts and seeds, participants dropped LDL cholesterol 33 percent.[16] For protection from all types of heart disease, a vegan diet with the inclusion of raw nuts and seeds is simply a healthier diet. However, the limitation of grains (especially flour) and the inclusion of green vegetables and low-starch vegetables in place of grains and starchy vegetables both play a role in the dramatic lipid-lowering benefits and weight loss.

When the diabetic diet is carefully designed around green vegetables and beans, with the addition of a small amount of fruit and a small amount of raw nuts and seeds, patients are able to stop insulin and sulfonylureas—the primary offenders that restrict weight loss—as soon as possible.

It is even more important for children, people who are thin, people who exercise a lot, and women who are pregnant and nursing to consume sufficient fat. The healthiest diet for all ages is one that includes some healthy fatty foods. This same diet will also prevent and reverse disease. There is no need for people with heart disease or diabetes to move to a special type of extremely low-fat vegan diet void of raw seeds and nuts, thinking this restriction is necessary or valuable for their cardiac health.

> For an overweight diabetic, I recommend one ounce daily of raw, unsalted seeds or nuts, such as sesame seeds, sunflower seeds, flax seeds, chia seeds, hemp seeds, pumpkin seeds, walnuts, pistachio nuts, or almonds. Add them to a salad or blend them into dressings as an oil replacement.

We are beginning to get a clear picture of how we can prevent and reverse diabetes. As the weight comes off, it is important to remember that the main risk of weight loss is weight regain. Dietary improvements that are not maintained are of no long-term benefit, and rapid weight gain is, of course, unfavorable. When you really "Eat to Live" you enjoy the combination of dramatic results for your body weight and your health, and your tastebuds get healthier too. We marry great flavor with the emotional satisfaction of knowing you are doing the best thing for your health and it becomes the way you prefer to eat for your life.

The other risk from weight loss is the formation of gallstones, or cholelithiasis. My experience, however, is that people losing weight following my dietary recommendations have an extremely low rate of gallstone formation. Certainly, the formation of gallstones and the possibility of laparoscopic gallbladder removal may be a reasonable price to pay for losing 40 to 140 pounds of life-threatening body fat. Nevertheless, it is important to note that the inclusion of raw nuts and seeds in the diet, especially while losing weight, is crucial protection against gallstone formation. Nuts are rich in several compounds that protect against gallstone disease.

The *American Journal of Clinical Nutrition* reported that when 80,718 women from the Nurses' Health Study, age thirty to fifty-five years in 1980, were followed for twenty years, researchers found that the consumption of nuts and seeds offered dramatic protection against gallstone formation. Women who consumed five ounces of nuts per week had a dramatically lower risk than did women who rarely consumed them. Further adjustment for fat consumption (saturated, trans, polyunsaturated, and monounsaturated fats) did not alter the relation.[17] These findings were also duplicated in a cohort of men.[18]

Perhaps, then, the reason I have not observed a high rate of gallstone formation in spite of having thousands of clients and patients all over the country who've lost large amounts of weight is the inclusion of nuts and seeds in the program. Because it is difficult to determine who might be at a higher risk for gallstone formation, it is wise for every dieter to include at least one ounce of raw nuts and seeds per day in any weight-loss program to offset this real risk to healthy weight loss.

Fat Deficiency Can Cause a Failure to Thrive

For many people, the undue emphasis on extremely low-fat diets has resulted in health difficulties. I have encountered many individuals who have not thrived on vegan or flexitarian diets. They may have developed dry skin, thinning hair, muscle cramps, poor sleep, and poor exercise tolerance. Often they do not realize their real problem. They go back to eating large amounts of animal products, not knowing that they were fat deficient on their low-fat vegan diet. For most of these individuals, eating more healthy fats from nuts and seeds, taking a

DHA supplement, and eating fewer starchy carbohydrates clears up the problem. This is not so uncommon. Some people simply require more essential fatty acids, both omega-6 and omega-3. DHA and EPA are those healthful, long-chain omega-3 fatty acids found in fatty fish and commonly known as fish oil. They are available from vegan sources today, mostly from algae or yeast.

A high-starch, low-fat diet—regardless of whether or not the dieter is eating meat—can derail weight loss and lead to high triglycerides, preventing lowered cholesterol levels. I have cared for some patients who came to me after they developed irregular heartbeats or cardiac arrhythmias. These conditions resolved when I added nuts and seeds back to their diet. Insufficient fat in the diet can also compromise the absorption of fat-soluble vitamins and healthful phytochemicals. When you eat a nut- or seed-based dressing on a salad, you absorb more of the carotenoids in the raw vegetables. More than ten times as much of certain nutrients are absorbed. A study detecting blood levels of alpha-carotene, beta-carotene, and lycopene found negligible levels after ingestion of salads with fat-free salad dressing but high levels after the same foods were eaten with fatty dressings.[19]

Ultimately, the nut icing on the carrot cake was displayed in the Adventist Health Study-1, a twelve-year study of thirty-four thousand Adventists in California. This group is the longest-lived population that has ever been formally studied in depth. We knew that the vegan and near-vegan Adventists lived longer than other Californians, but what were the precise factors accounting for the years of life gained? Interestingly, this study found that the strongest effect of any food on promoting longevity was the consumption of nuts or seeds five or more times per week. The consumption of nuts added years of life, likely due to the antiarrythmic effect of nuts and seeds, compared to non–nut eaters, who suffered double the rate of fatal coronary events.[20] For any population at risk of heart disease, especially diabetics, eating some seeds and nuts daily is imperative and maybe even lifesaving.

Let's take a look at a day's menu of equal calories both with and without nuts and seeds, to see some subtle but important nutritional differences.

WITHOUT NUTS/SEEDS	WITH NUTS/SEEDS
Breakfast	
Oatmeal	Oatmeal
Blueberries and Dates	Blueberries, Walnuts, Flax Seeds
Lunch	
Salad with Fat-free Italian Dressing	Salad with Caesar Salad Dressing/Dip*
Baked Potato with Broccoli	Broccoli with Red Lentil Sauce*
Whole Grain Bread	Apple
Dinner	
Raw Veggies with Fat-free Ranch Dressing	Raw Veggies with Garbanzo Guacamole*
Easy Bean and Vegetable Chili*	Easy Bean and Vegetable Chili*
Brown Rice	Fruit Bowl
Whole Wheat Roll	

Note: * indicates recipe included in this book.

	WITHOUT NUTS/SEEDS	WITH NUTS/SEEDS
Total Calories	1,882	1,878
Fat	21 gm 9.2 %	66 gm 28 %
Carbohydrate	381 gm 76.8 %	277 gm 54 %
Protein	69.5 gm 14 %	85.5 gm 18 %
Arginine (amino acid)	3,627 mg	5,806 mg
Vitamin E	.29 mg	.66 mg
Sodium	1,570 mg	644 mg
Calcium	978 mg	1,356 mg
Iron	24 mg	29 mg
Phosphorus	1,387 mg	1,694 mg
Magnesium	540 mg	750 mg
Zinc	9.6 mg	13.6 mg
Copper	2.2 mg	4.6 mg

You can see that the diet with nuts and seeds is higher in protein and much higher in the amino acid arginine. Arginine has special properties that benefit the heart, promoting vasodilatation (relaxation of the vessel wall) and preventing blood clotting. It also includes higher amounts of vitamin E and minerals, but that does not adequately reflect the major difference between these diets. The very low-fat diet greatly reduces the absorption of most of the carotenoids and other phytochemicals contained in a meal. They simply are not well absorbed in such a low-fat environment. The benefits of nuts and seeds are enhanced when they are eaten with a meal; they are not for snacking.

There is one more fascinating gem about nuts and seeds, and I do not want you to forget it: the calories are not all biologically available. They are similar to the resistant starch calories in beans. About 30 percent of the recorded calories from nuts and seeds are passed into the stool, not absorbed into the bloodstream.[21] Eating nuts and seeds increases stool fat, which means not all the fat is absorbed. Plus, the sterols, stanols, and other sponge-like fibers in nuts and seeds carry other calories from the diet into the stool as well. So in the calculated dietary menus above—one with nuts and seeds, one without—even though the calories recorded and eaten are about the same, the amount of absorbed calories from the diet plan with the nuts and seeds would be about 100 calories lower. Interestingly, however, the increase in fecal fat and fiber does not occur when the diet contains the oils instead of whole nuts and seeds. In other words, calories from oil are absorbed almost 100 percent. For example, eating whole peanuts versus peanut oil would have completely different biological effects.

Remember, it is best to eat nuts and seeds raw or only lightly toasted. When you roast nuts and seeds, you form carcinogenic acrylamides as the food is browned. You decrease the protein and create more ash from the roasting process. The more the nuts and seeds are cooked, the more their amino acids are destroyed. You also lower levels of calcium, iron, selenium, and other minerals in the roasting process.

With your growing awareness of the health properties of nuts and seeds, please take into consideration that they should be eaten in moderation. Should we sit in front of our TVs, eat an entire bag of nuts in an hour, and complain when we gain weight? Of course not. Healthy eaters avoid excessive calories and do not eat for recreation. Eat only one ounce a day if you are overweight. If you are thin, physically active, pregnant, or nursing, eat two to four ounces daily according to your caloric needs.

As we are beginning to see clearly, ideal health has very little to do with a precise ratio of carbs, fats, and proteins. To prevent and reverse diabetes, we must make certain that we are paying attention to our nutrient-per-calorie model. We need to eat foods with micronutrients and other proven health benefits to achieve our health goals.

It's time for action. It's time to eat the right foods that lead to long-lasting health.

CHAPTER EIGHT

The Nutritarian Diet in Action

I am one hundred pounds overweight. Three weeks ago, I had a physical and my blood sugar was 289. The doctor told me I was diabetic and would be on medication for the rest of my life. Instead of taking the medication the doctor prescribed, I started your program on Saturday, May 19. Three days later, my glucose was 90 points lower! Every day it is getting lower. My fasting glucose is now 117, and when I check at night before going to sleep it is 120 to 122. A drastic change from 289—in less than three weeks' time! I will continue this plan and my exercise regimen so that my numbers get even lower. I am losing about five pounds a week. Your advice has saved my life. Thank you, Dr. Fuhrman.

—Laverne Stone, age fifty

For diabetics, optimal health is not achieved by eating less to lose weight. Rather, diabetics have to make a big commitment not only to forming better eating habits but also to eating the *right foods* to help the body heal. This change includes making adjustments in their thinking as well as their diet. But it can be challenging to change your thinking when the doctors and the media are spewing misguided information.

A recent article published by the Associated Press quotes a former president of the ADA: "There is no special diet [for diabetics to lose weight]. You have to eat fewer calories than your body burns," said Dr. Robert Rizza, a Mayo Clinic endocrinologist and former president of the American Diabetes Association.[1] Nothing can be more untrue. A diabetic's life or death can most certainly depend on the quality and not just the quantity of what is eaten. It's essential to understand *what* to eat, not just how many calories are in the doughnut or large French fry order. A diabetic's life depends on it. All of us want a healthy life expectancy. Who would want to suffer needlessly in their later years?

High-quality foods that are rich in micronutrients are your body's best friend. They fuel your body's self-repair mechanisms; they curtail your appetite; and they are the ticket to a slim, healthy body free of diabetes and heart disease. The right diet style can remove your addictive drive to overeat and enable you to control your chronic overeating for the first time. It can save your life.

When you eat a nutrient-rich diet, you are eating more food volume, more food by weight, and more high-water-content food, meaning you may feel more full after a meal even though you are eating fewer calories overall and less food that has a high caloric concentration.

High-Nutrient, High-Volume Foods to Fill Up On

I call high-nutrient, high-volume foods that fill you up the unlimited foods. They include:

1. All raw vegetables

2. All cooked green vegetables

3. Tomatoes, peppers, eggplant, mushrooms, onions, cauliflower

The goal is to eat large amounts of these three food categories to flood the body with micronutrients and fiber. Memorize them!

So these foods richest in nutrients per calorie are also naturally low in calories. You get more nutrients and fewer calories simultaneously. That is the real fountain of youth.

Fruit is not unlimited, but you can eat a few with breakfast and one with lunch and dinner if desired. Beans are not totally unlimited, but they are a recommended food that you should eat liberally each day. You can eat lots of them, up to a cup of beans with each lunch and dinner, two cups total a day. Nuts and seeds are an important and healthy food to include in your diet, but they should be limited to one ounce a day for overweight females or 1.5 ounces a day for overweight males. Obviously, these limits can be liberalized for people who are not overweight or diabetic and require more calories.

Remember, raw vegetables and all cooked green and non-starchy vegetables such as string beans, artichokes, zucchini, snow peas, eggplant, tomatoes, peppers, mushrooms, cauliflower, onions, and leeks do not have to be measured. Eat as much as you like of these dishes made with gentle spices, tomatoes, onion, and garlic.

The Salad Is the Main Dish

Raw vegetables should be eaten in large quantities at the beginning of each main meal. A sensible goal is to shoot for one pound of raw vegetables daily. I often say, "The salad is the main dish." The word *salad* here means any uncooked vegetable. Think big when it comes to salad. The more raw vegetables you eat, the less you will desire of everything else. Raw vegetables are the healthiest, most weight-loss-promoting foods you can eat. Use a variety of raw vegetables in your salads. In addition to plenty of lettuce, include tomatoes, shredded beets, carrots, cucumbers, and peppers. You can also add any leftover steamed greens on top of a lettuce salad, some defrosted frozen peas, stewed mushrooms, or a handful of beans. Add a little fruit-flavored vinegar or one of my dips or dressings, and the salad can be the whole meal.

RECOMMENDED SALAD VEGETABLES

Lettuces—all varieties

Tomatoes	Celery	Zucchini
Carrots	Broccoli	Onions and Scallions
Radishes	Cauliflower	Sprouts
Fennel	Baby Bok Choy	Cucumber
String Beans	Snap Peas	Snow Peas
English Peas	Endive	Peppers
Hearts of Palm	Water Chestnuts	Stewed Mushrooms (chilled)*

The increased production of the biologically active phytochemicals in raw vegetables is consistent with the studies that show a radically lower risk of cancer in people who consume more raw greens in their diet.[2] For those in the know, eating lots of raw greens is the most important nutritional intervention to prevent common human cancers.[3] Eat not only lots of raw greens every day but also big portions of cooked green vegetables. Remember, if it's a vegetable with the color green, it is rich in micronutrients and low in calories. It's a green light to eat more of it. The more greens you eat, the increased likelihood you will eat less of something else that is higher in calories. When you fill up with high-volume, high-nutrient foods that have a high micronutrient content, you will not feel the need to overeat foods that sabotage your health and weight. The added benefits include protection against heart attacks and cancer. Of course, green vegetables are the food that shows the most protection from diabetes too.[4] How great is that?

---

*Mushrooms are best cooked, even stewed, for a few minutes. They contain a mild toxin called agaritine, that dissipates with even light cooking. To be conservative, because it is unknown if agaratine has negative health effects, I do not recommend eating much raw mushrooms. Nevertheless, mushrooms have powerful anticancer effects, and those powers are likely enhanced by cooking them.

With the growing popularity of nutritional supplements, more and more Americans are looking for accurate information about nutrients that can make a real difference in their health and their lives. However, the reality is that the most powerful thing people can do to improve their health is eat more green vegetables. Americans eat a piddly amount of greens, but if they ate much more, disease rates of all types would plummet. Not only are vegetables rich in discovered vitamins and minerals, but as discussed, they also contain thousands of phytochemicals that are critically important to our health.

The more we get a better understanding of nutritional science, the more we learn that individual nutrients taken as supplements do not have the same healing and protective powers of high-nutrient, superfoods. Supplements can be used to supplement an area of potential suboptimal intake, such as B12, iodine, zinc, vitamin D, or DHA, but they never can take the place of eating healthfully. Not all vegetables are created equal, and one of the most fascinating areas of research in the last ten years has been the therapeutic value of cruciferous green vegetables.

The World's Healthiest Foods

Almost three hundred case-control studies show that vegetable consumption provides a protective effect against cancer and that cruciferous vegetables are the foods with the most powerful anticancer effects of all foods. Cruciferous vegetables are those in the broccoli and cabbage family and include such foods as bok choy, radishes, and watercress. While eating fresh fruits, beans, vegetables, seeds, and nuts have been all been shown in scientific studies to reduce the occurrence of cancer, cruciferous vegetables are different. Instead of a one-to-one relationship against cancer, they have a one-to-two relationship with a wide variety of human cancers. In other words, as plant food intake goes up 20 percent in a population, cancer rates typically drop 20 percent. But as cruciferous vegetable intake goes up 20 percent, cancer rates typically drop 40 percent.[5]

Cruciferous vegetables contain phytochemicals that have a unique ability to modify human hormones, detoxify compounds, and prevent toxic compounds from binding to human DNA, thus limiting toxins from causing DNA damage that could lead to cancer. Sulforaphane, an extensively studied compound, is an isothiocyanate that has a unique mechanism of action. This compound blocks chemical-initiated tumor formation and induces cell cycle arrest in abnormal cells, meaning it inhibits growth and induces cell death in cells with early cancerous changes in a dose-dependent manner. Recent studies show that the amount of sulforaphane that can be obtained from eating a reasonable amount of broccoli can have dramatic effects to protect against colon cancer.[6]

CRUCIFEROUS VEGETABLES

Kale, collards, broccoli, broccoli rabe, brocollina, Brussels sprouts, watercress, bok choy, cabbage, Chinese cabbage, mustard greens, arugula, kohlrabi, red cabbage, mache, turnip greens, horseradish, rutabaga, turnips, radishes

Most micronutrients that we read about (vitamin C, vitamin E, beta-carotene, lutein, lycopene) function as antioxidants in the body, meaning these compounds neutralize free radicals, rendering them harmless. But the phytochemicals in cruciferous vegetables do more. They enable the body's own antioxidant control system. When we take in direct antioxidants such as vitamin C and E, they fight a one-on-one war against free radicals but not much more. Their effects are gone in a few hours. Synthetic or isolated fractions of vitamin E, beta-carotene, or vitamin C can also cause pro-oxidant behavior, creating more free radicals. So instead of having a short-lived benefit that gets used up quickly, the unique compounds in cruciferous vegetables cycle over and over, protecting the body for three to five days after consumption. They fuel body systems already in place so that they function more effectively, but they also induce many different systems to defend against many different types of damage.

CRUCIFEROUS VEGETABLES
- Repair and protect
- Detoxify toxins and carcinogens, rendering them harmless
- Regulate the liver's ability to remove toxins
- Remove free radicals to prevent oxidative and DNA damage in cells
- Transform hormones into beneficial compounds that inhibit hormone-sensitive cancers
- Enhance and protect against age-related loss of cellular glutathione
- Enable cell death in cells that have abnormal mutations and DNA damage[7]

A study on prostate cancer shows that it takes twenty-eight servings of vegetables a week to decrease the risk of prostate cancer by 33 percent, but just three servings of cruciferous vegetables a week decreases the risk of prostate cancer by 41 percent.[8] The National Cancer Institute of the National Institutes for Health recommends nine servings of fruit and vegetables per day. I recommend eight servings of vegetables a day, with at least two of those being cruciferous vegetables, one raw and one cooked. Do you eat green cruciferous vegetables daily? Do you eat both raw and cooked green vegetables? Let's all make sure we do.

Greens and Heart Disease

Heart disease is caused by the buildup of fatty plaques in the arteries, known as atherosclerosis. However, arteries do not get clogged up with these plaques in a uniform way. Bends and branches of blood vessels—where blood flow is disrupted and can be sluggish—are much more prone to the buildup. A recent study shows that Nrf2, a protein that usually protects against plaque buildup, is inactive in areas of arteries that are prone to disease.[9] However, a phytochemical found in green vegetables activates Nrf2 in these disease-prone regions. Activation of Nrf2 is important for maximizing both prevention and removal of plaque. Ingestion of these beneficial compounds from cruciferous green vegetables had the strongest effect to activate the Nrf2 proteins, blocking atherosclerosis.

Vegetables have other powerful disease-fighting nutrients as well. Carotenoids are just one compound that is important for excellent health. Greens have high levels of carotenoids and other nutrients that prevent age-related diseases that diabetics are prone to. For example, the leading cause of age-related blindness in America is macular degeneration. If you eat these greens at least five times per week, your risk drops by more than 86 percent.[10] Lutein and zeaxanthin are carotenoids found in green vegetables with powerful disease-prevention properties. Researchers have found that people with the highest blood levels of lutein had the healthiest blood vessels, with little or no atherosclerosis.[11]

Of course, losing body fat, lowering glucose levels, and lowering blood pressure all play an important role in the prevention and reversal of heart disease. But the message here is that a diet rich in greens protects against diabetes, aids in weight loss, and fights cancer, in addition to enabling more targeted and effective reversal of high blood pressure and heart disease.

And green vegetables are so incredibly low in calories and rich in nutrients and fiber that the more you eat of them, the more weight you will lose. I am going to stop here and highlight this point.

The more green vegetables you eat, the
more weight you will lose.

This is an incredible gift of nature. One of my secrets to nutritional excellence and superior health is the one pound rule. That is, try to eat at least one pound of green vegetables a day, combining your raw and cooked greens.

This may appear to be an ambitious goal at first, but I assure you, you will achieve the dietary balance and results you envision when you work toward it. Let's say it again: the more greens you eat, the more weight you will lose. The high volume of greens will not only be your secret to a thin waistline, but it will also simultaneously earn you future protection against life-threatening illnesses.

And lastly, while everyone jumps on the cruciferous-vegetables-are-good-for-you bandwagon and more science continues to build on their powerful health benefits, let's not forget H = N/C. In other words, besides all these unique features, green cruciferous vegetables still contain more vitamins and minerals per calorie than any other food on the planet.

Note that green vegetables are relatively high in protein per calorie.

For many years, most Americans incorrectly believed that only animal products contained all the essential amino acids and that plant proteins were incomplete. False. They were taught that animal protein is superior to plant protein. False. They accept the outdated notion that plant protein must be mixed and matched in some complicated way that takes the planning of a nuclear physicist for a vegetarian diet to be adequate. False.

I guess they never thought too hard about how a rhinoceros, hippopotamus, gorilla, giraffe, or elephant could become so big while eating only vegetables. Animals do not make amino acids from air; all the amino acids originate in plants. Even the nonessential amino acids that are fabricated by the body are just basic amino acids that are modified slightly by the body. So the lion's muscles can only be composed of the protein precursors and amino acids that the zebra and the gazelle ate. Green grasses (or leafy greens) made the lion and are the mother of all the protein that built all the creatures on planet Earth.

I've asked hundreds of people this question: which has more protein—one hundred calories of sirloin steak or one hundred calories of broccoli? The typical answer is, "Steak, of course." When I tell them it's broccoli, the most frequent response I get is, "I didn't know broccoli had protein in it." I then ask, "So where did you think the calories in broccoli come from? Did you think it was mostly fat, like an avocado, or mostly carbohydrates, like a potato?"

People seem to know less about nutrition than any other subject. Even the physicians and dietitians who attend my lectures quickly volunteer the answer, "Steak!" They are surprised to learn that broccoli has more protein per calorie than many cuts of meat and that if you eat large quantities of green vegetables, you receive a considerable amount of protein. One ten-ounce bag or box of frozen broccoli contains over 10 grams of protein. When you get most of your protein from greens and beans, you get a bonanza of protective health benefits in the process—not to mention that fountain of youth people have been searching for.

PROTEIN CONTENT OF SELECTED PLANT FOODS (IN GRAMS)	
Spinach (frozen, 1 cup)	7
Collards (2 cups)	8
Peas (frozen, 1 cup)	9
Almonds (3 ounces)	10
Broccoli (2 cups)	10
Tofu (4 ounces)	11
Sesame Seeds (½ cup)	12
Kidney Beans (1 cup)	13
Sunflower Seeds (½ cup)	13
Chick Peas (1 cup)	15
Lentils (1 cup)	18
Soybeans (1 cup)	29

The Eating Plan

Think about setting up your menu plan in two phases. The first is a more strict phase in the first few weeks to get your weight down to a safer number and to reduce as much of your medication as possible early on. In this phase, other than a limited amount of beans, avoid high-carb foods and eat no animal products. Of course, you can remain in phase one as long as you need or want to maximize results. What follows here are general guidelines to understand phase two of this program. The main difference between these guidelines and phase one (when you are looking to control your out-of-control glucose numbers and reduce medications) is more restriction on fruit, grains, and starchy vegetables. The exact specifics of the phase one plan will be laid out in chapter 10.

Then move on to phase two, a more livable phase, once you have reached a stable place. Then you could add one serving a day of peas, butternut or acorn squash, and intact whole grains such as black wild rice or steel-cut oats. If you're eating animal products, try to do so only in very small amounts, such as a condiment to flavor a vegetable dish or soup, so it is not a major calorie source. Use only about one ounce for this purpose, and do not use more than two ounces in one day.

Breakfast

Breakfast should consist of a few low-sugar fruits, such as berries, papaya, kiwis, pomegranates, oranges, and green apples. You can eat a whole orange or apple, but do not drink orange or apple juice. Limit fruits to a total of five servings daily, usually two to three fresh fruits with breakfast, one after lunch, and one after dinner. Blueberries, raspberries, blackberries, and strawberries are especially recommended because of their concentration of important phytochemicals and overall high ANDI score. Consider using frozen berries when fresh are not available. Because berries are expensive, many people go to a berry-picking farm in the summer and buy large amounts at a great price. You can save money and get high-quality food when you pick your own and then freeze them in serving-sized bags for use all winter. Eat these fruits with some raw vegetables such as cucumbers, celery, fennel, or lettuce.

A half cup of cooked oats—or even a whole cup for men—is also acceptable to eat with breakfast. Use rolled oats or steel-cut oats (preferred) but not quick oats. Raw oats have more resistant starch than cooked oats, so try soaking steel-cut oats and regular rolled oats overnight in the fridge and then eating them soft, but not cooked, the next morning. You can put fruit on top too. Another great option is a squash-based breakfast soup, made with cooked butternut squash, chopped greens, almond milk (almonds and hemp seeds blended with water), cinnamon, nutmeg, and chopped apple. Baked eggplant chopped with onion, shredded green apple, chia seeds, crushed walnuts, cinnamon, and nutmeg also makes a delicious breakfast dish.

Try to eat one tablespoon of ground chia seeds or flaxseeds daily with breakfast. Sprinkle them over a bowl of cut fruit or mix them into the oats. Ground chia and flaxseeds are rich in omega-3 fatty acids, lignans, and other beneficial fibers, so they are strongly recommended. These seeds have protective effects for the heart and against cancer. Also, eat four walnut halves every day with breakfast. They also have highly protective properties. A mixture of sliced apples, flax or chia seeds, crushed walnuts, just a sprinkle of raw oats, and cinnamon makes a great breakfast treat.

Lunch

Lunch is the most important meal of the day because people are usually away from their home, in public, at work or school, and around others. When you're away from your own kitchen, make sure to bring plenty of good-tasting and filling food to keep you satisfied. When you have a lunch with you that you enjoy, it makes the day pleasant and you'll find it easier to stick to your healthy eating plan.

Make salad your main lunch dish. Always include lots of raw vegetables. Dressings or dips can be made using raw ground nuts and seeds or an avocado as the base. Top your salad with some beans and a great-tasting dressing, and finish with a piece of fruit for dessert. Delicious dressing and dip recipes are available in the menu section of this book with many more options on my member website at www.drfuhrman.com.

One serving of any type of fruit, not just the low-calorie ones, is okay because the sugar load will be diluted with the rest of the vegetable meal. A mango, peach, pear, orange, or banana are good ideas. Or you can have a salad and a bowl of vegetable-bean soup and a piece of fruit for dessert.

Prepare a huge pot of vegetable soup on the weekends. Portion it into separate small containers so you can just grab a container out of the refrigerator and drop it in your lunch bag with some raw greens, tomato, and other raw veggies during your busy week. Be sure to include a small container of your favorite healthy salad dressing. Eat the soup at room temperature or warm it up at work for a hearty and filling meal. Remember, this is the beans-and-greens diet, or the salad-and-soup diet, because of the abundance of greens in your salad and green vegetables with beans in your soups. Sometimes I like to pour the hot soup over my shredded raw salad vegetables and use the soup as a heavy dressing for the lettuce, spinach, and cabbage that I have shredded very thin.

I like to make salad dressings on Sunday and Wednesday so they stay fresh. The Sunday dressing lasts me for three days, and the Wednesday dressing lasts for four days. Measure out the ounces of nuts and seeds (or avocado) you're using in the recipe so you can then divide the full recipe according to the appropriate servings of nuts per day. So, if you are making the dressing for one person and you include four ounces of seeds and nuts, separate that dressing into four separate containers for use over the next four days. Your flaxseeds and four walnut halves with breakfast do not count in the one ounce of permitted seeds and nuts per day.

Dinner

Dinner begins with another salad, or raw vegetables with dip. A tomato-based salsa-type dip or hummus are the most favored choices. Try the low-calorie, high-nutrient dips and dressings featured in this book. If you used the soup as your salad dressing at lunch, use the nut- or seed-based salad dressing with dinner.

Always have a large plateful of steamed green vegetables with dinner. Choose from a wide variety of vegetables such as string beans, cabbage, bok choy, artichokes, brussels sprouts, kale, broccoli, and zucchini. Frozen vegetables can be used instead of fresh. The prototype diet is a large salad with a bowl of vegetable bean soup at lunch and a large salad with steamed greens at dinner, but you can switch the meals around. And if you're not having soup with your meal, be sure to include some beans, either a bean dish or unsalted canned beans tossed on top of your salad.

A vegetable stew such as a ratatouille made with peppers, onions, tomatoes, and zucchini, and herbs or some grilled vegetables are options to eat after your dinner salad. A bean burger with sliced raw onion and tomato is another option, or if it is your fish night, have the salad, steamed greens, and a small piece of fish cooked with garlic, onions, and tomato.

Or eat a light dinner of salad and greens cooked with onions and mushrooms, with some beans scooped on top, and finish it off with a delectable sorbet for dessert. Frozen fruit whipped in a high-powered blender with some fresh fruit makes a delicious sorbet dessert treat. Or have a slice of melon or a bowl or berries.

You have now come to the end of your food allowance for the day. A good habit to adopt is to floss and clean your teeth at this point as a reminder to stay out of the kitchen and away from food the rest of the night.

> ## Steamed Artichokes Are Luscious
>
> Slice one inch off the top of an artichoke. Cut off the very bottom of the stem, but keep the rest of the stem attached—it's edible and delicious. Slice the artichoke in half lengthwise. Use a sharp knife to slice out the fuzzy section in the center of the choke. Remove the triangle of small leaves that cover the fuzzy center too. Now the artichoke will cook much faster than it would if it were whole. Steam it for eighteen minutes, until the outer leaves pull off easily. Drain and serve.
>
> You can eat the heart stem and most of the smaller leaves as well as the tender insides of the larger leaves. Artichokes have such a delicious, subtle flavor, they require no seasoning.

Other General Guidelines

Starchy vegetables should be limited in the first phase of this diabetes-reversal program. For maximizing weight loss, the trick is to use beans as your primary starch source and only one other serving a day of a non-bean starchy food like beets, carrots, peas, and squashes. If you're eating a one-cup serving of a starchy whole grain, such as oatmeal, steel-cut oats, or wild rice with breakfast, do not eat the starchy vegetable option with dinner.

The amount of starchy vegetables varies with body weight and exercise habits. Slim people who do not have diabetes and exercise a lot can have more starches, and more seeds and nuts, to meet their higher caloric needs. For overweight diabetics, these are more limited foods. Colorful cooked starchy vegetables such as beets, carrots, corn, and butternut and winter squashes can be eaten in small amounts with dinner. High-starch foods made from flour, rice, or white potatoes are even more limited in the recommended diabetic menu. It's preferable to eliminate them from your diet until you're completely off insulin and sulfonylureas.

Phase one of this program is for people coming off insulin and other diabetes medications, so that occurs as quickly as possible, because in most cases, the more medications you require, the more weight loss is hindered. If you follow your blood sugar closely, with physician monitoring, the medications can be reduced and in many cases eliminated. During phase one, I recommend no high-starch vegetables or grains. Instead it is better to get your carbohydrate and calorie requirements from cauliflower and beans. That way you will be utilizing only the starchy foods that have a low GL and that are high in resistant starch. Make use of the non-starchy cooked vegetables and tofu to prepare filling, low-calorie stews. Eggplant, tomatoes, onions, garlic, mushrooms, peppers, beans, and zucchini form the base of these dishes.

No sweetened drinks of any type are permitted, even artificially sweetened. Even no-calorie sweeteners can stimulate the pancreas to work. No fruit juice. Vegetable juice can be used as part of the soup base—dilute it with water. Drinking only water and eating whole foods are strongly recommended. In general, drinking our calories is unfavorable for diabetics.

Dried fruit, such as raisins, are limited to a minimal amount, usually only as a flavor enhancer as a small part of the recipe in a breakfast dish, soup, or vegetable dish.

Seeds and nuts are limited to one to two ounces daily, depending on weight and activity level—usually a one-ounce limit for overweight women and a 1.5 ounce limit for overweight males. Seeds like raw sunflower, chia, hemp, raw unhulled sesame, and pumpkin seeds are great choices, and even preferred over raw nuts, as they are higher in nutrients and have beneficial fatty acid profiles. A half of an avocado is permitted occasionally on a salad or with a dip, but make sure the seeds and nuts do not exceed one ounce when you're also using the half avocado.

Refined flour products, bread, white rice, processed cold cereals, and white potatoes are not allowed in these menus, as these foods are not recommended on a regular basis. Nor do these menus and recipes contain added salt, oil, or sweeteners of any kind.

Whole milk, cheese, butter, and red meat are not recommended. These foods should only be considered on special occasions or holidays. Nonfat dairy products could be used as a flavoring in small amounts once or, occasionally, twice a week. However, they are not missed and can easily be replaced in recipes with almond or hemp milk and other nondairy alternatives. Blending a half cup of raw almonds and hulled hemp seeds with three cups of water in a high-powered blender makes a simple milk when a recipe calls for that.

Animal products in general should be limited to a small amount of fish once a week and then only one other small serving of non-fish white meat per week. No eggs, meat, or cheese.

Usually, you can still achieve good results with a small amount of animal products such as one ounce of turkey, scallops, shrimp, or chicken to flavor a soup, stew, or stir-fry. Many people feel more satisfied when they are allowed to have even this small amount of their animal product allotment divided up as a condiment to flavor dishes. Instead of eating one four-ounce serving a week, they split it up over several meals.

Some recipes in my menu plans are used twice in the same day or as leftovers the next day. This is done intentionally because when you prepare a dish, it makes sense to reduce your workload and make enough for at least two meals. Many people choose to use a prepared dish for two to three days to save time and cooking efforts. My family cooks huge amounts of food and eats leftovers for a few days. That way we only have to prepare food two days a week. Experiment to find out what works best for you.

With the removal of potato, rice, and flour products, and the restriction on nuts and seeds to one ounce daily, you may need to add more beans to reach your caloric needs. Greens, beans, fruit, nuts, seeds, tofu, and the low-calorie vegetables listed above supply the major volume of calories in the diet. Nuts and fruits somewhat less, of course. Spaghetti squash and cauliflower are permissible substitutions for higher-starch, higher-calorie grains and potatoes. Turnips, parsnips, and other squashes are alternatives to rice, bread, and potatoes and are a better, more nutrient- and fiber-rich choice. Grains should be whole and intact when cooked in water. Whole grains such as brown and black rice, barley, quinoa, and steel-cut oats or old-fashioned oatmeal are ideal examples; nevertheless, they are best avoided in phase one and limited to a one-cup serving or less per day in phase two. Most of my preferred diabetic-reversal menus have such grains only a few times per week.

Always wash fresh fruit and vegetables thoroughly. Buy organic if possible. Always buy organic strawberries, spinach, and celery, as these three items are the most pesticide-contaminated foods in the produce section.

Cooking Techniques and Tips of the Trade

A basic cooking technique utilized in some of these recipes is water-sautéing. This is used instead of cooking with oil. Water-sautéing is simple and good for stir-fries, sauces, and many other dishes. To water-sauté, heat a skillet on high heat until water sputters when dropped in the pan. Use small amounts of water, starting with two to three tablespoons in a hot skillet, wok, or pan, then adding the finely sliced vegetables, stirring and then covering to maintain the moisture. Continue to stir and add more water only if necessary. In many dishes, the moisture from tomatoes, mushrooms, zucchini, and other high-water-content vegetables is sufficient.

Soups and stews are critical components of this nutritarian diet style. When vegetables are simmered in soup, all the nutrients are retained in the liquid. Many of the soup recipes use fresh vegetable juices, especially tomato, celery, and carrot juice. These juices provide a very tasty antioxidant-rich base. If you don't have a juicer, consider purchasing one.

If you are short on time, bottled tomato and other vegetable juices can be purchased at most health food stores, but nothing beats the flavor of freshly juiced vegetables. I also use a simple procedure to create "cream" soups. Raw cashews or cashew butter are blended into the soup to provide a creamy texture and rich flavor. A big advantage of homemade soups is that they make wonderful leftovers. Soups generally keep well for up to four days in the refrigerator but should be frozen if longer storage is desired.

Should you occasionally choose to use a prepared soup, keep in mind that your overall daily sodium intake should remain under 1,200 milligrams for men and under 1,000 milligrams for women. Natural whole foods contain 400 to 700 milligrams of sodium, which allows for a leeway of about 500 milligrams. Be sure to read labels. You will be amazed by how much sodium canned soup contains. Try to select a no-salt added variety.

My recipes include many delicious salad dressings and dips. Conventional dressings usually start with oil and vinegar; the oil provides the fat, and the vinegar provides the acidity. The fat sources in my salad dressings are whole foods such as raw almonds and cashews, other raw nuts and seeds, avocado, and tahini. This is not a fat-free eating style because our bodies require healthy fats from whole foods; the way nature designed us to consume them. By eating this way, we receive the lignans, flavonoids, antioxidants, minerals, and other protective phytochemicals that come along in the package. So the oil is removed, and seeds and nuts supply the healthy fats instead because they are such a healthy disease-fighting food. Removing the oil and using nuts and seeds as the primary fat source in the diet is critical to reversing diabetes.

A powerful blender such as a Vitamix is very helpful for making salad dressings, creamy soups, smoothies, and fruit sorbets. Nuts and seeds do not get soft and creamy for dressings in a regular blender unless more liquid is added. Only the more expensive, high-powered blenders can make fruit sorbets from frozen fruits and blend vegetables effortlessly for fruit and green smoothies.

Frequent Mini-Meals All Day or Three Meals a Day?

Here is a quiz to see if you understand part of this complicated message. Is it healthier to eat frequent small meals or just two or three meals a day?

I'll give you a hint. Remember, deep cellular repair and detoxification occurs most readily in the non-digestive stage (during glycolysis) and during sleep. I am sure you now have the answer.

Contrary to popular advice, it is best to eat only when you're hungry, and not eat after dinner. Your next meal should be when you feel true hunger the following morning. Giving the pancreas the prolonged time to rest and to lower insulin levels is key. Snacking after finishing dinner is the worst thing you can do. When you consider the advice to eat frequent small meals and snack all day long, remember this:

1. People who eat more frequently usually take in more calories per day.

2. Obese people are invariably snackers; snacking correlates with obesity.

3. Animals fed the same calorie amounts monthly, but on a less-frequent schedule, live longer.

4. Diabetic recovery is enhanced by resting, not utilizing, the pancreas.

5. You don't want or need to eat frequently when you eat properly. Food tastes better when you wait until true hunger occurs.

6. Eating when truly hungry directs you to your precise calorie needs to maintain a lean body. Eating outside of true hunger

is usually addictive eating or recreational eating and promotes diabetes.

7. More frequent eating or snacking has been shown to increase the risk of colon cancer in men.[12]

The Diabetes Solution: High-Nutrient, High-Fiber, High-Water-Content Foods, with No Snacking

Most people following this program for diabetes reversal eat two main meals of mostly vegetables and beans plus a light breakfast. That means they usually eat a breakfast consisting of about 300 calories and then about 400 to 500 at lunch and again at dinner. Because type 1 diabetics are on insulin, they should eat approximately the same amount of calories at each meal and stay with three meals a day.

Some people claim that more frequent eating speeds up the metabolic rate, and a faster metabolic rate facilitates weight loss. This is a myth. Eating more frequently or eating more calories will not change metabolism sufficiently to make up for the increase in calories. If eating less made you fat, then anorexic people would be the heaviest. Of course, overweight people naturally have slow metabolic rates. That was a favorable genetic inheritance to enhance survival for most of human history when food was not as plentiful as it is today. A slow metabolism means a person can comfortably eat less and not get too thin. Actually, a slower metabolic rate means one is aging slower too. Therefore, if you are a naturally heavy person, you have been given a survival gift. A slower metabolic rate is only a bad thing in today's toxic food environment, not the food environment that has existed for most of human history. You can still have a long, healthy life free of diabetes and heart disease. It is just that you are designed efficiently and can handle lots of physical activity without needing too many calories.

People on this diet style are actually amazed how satisfied they are with fewer calories. Since they are getting so many micronutrients, they feel great, have more energy, and definitely are not suffering the same hunger cravings often felt on most restrictive diets. You can eat abundantly from healthy foods and still not overeat calories.

The only safe way to enhance metabolism is exercise. Supplements, green tea, caffeine, metabolic boosting herbs and supplements, drinking hot or cold drinks, or packing yourself in ice do not play a significant role in weight loss.

Supplements and Multivitamins to Consider or Avoid

Multivitamins and supplements have pros and cons. The main problem with taking a typical multivitamin is that it may expose you to extra nutrients that not only are unnecessary for your body but could actually be harmful too. Excessive quantities of some vitamins and minerals can be toxic or have long-term negative health effects. We know it is important to avoid vitamin and mineral deficiencies, but it is just as important to avoid consuming too much of certain nutrients.

Folate

Folate and folic acid are members of the B vitamin family. Folate is the form found naturally in foods, especially green vegetables and beans. Too much folate obtained naturally from food is not a concern. It comes naturally packaged in balance with other micronutrients, and the body regulates its absorption. Folic acid is the synthetic form that is added to food or used as an ingredient in vitamin supplements. Folic acid is also added to most enriched, refined grain products like bread, rice, and pasta in the United States and Canada in an attempt to replace the nutrients lost during the processing of the whole grain. Recently, there have been troubling studies connecting folic acid supplementation with increases in breast, prostate, and colorectal cancers.[13] A diet rich in green vegetables is high in folate, so supplemental folic acid is not necessary on this kind of diet. It is important for our health to eat vegetables to obtain the folate (and other nutrients) we need and avoid the significant risks associated with supplemental folic acid.

Vitamin A

Vitamin A is also risky to take in supplemental form. Ingesting vitamin A or beta-carotene from supplements instead of food has been shown to increase the risk of certain cancers.[14] In Finnish trials, using beta-carotene supplements failed to prevent lung cancer, and there was actually an increase in cancer in those who took the supplement. This study was halted when the physician researchers discovered the death rate from lung cancer was 28 percent higher among participants who had taken the high amounts of beta-carotene and vitamin A.[15] The death rate from heart disease was also 17 percent higher in those who had taken the supplements compared to those just given a placebo. Another recent study showed similar results correlating beta-carotene supplementation with an increased occurrence of prostate cancer.[16] Furthermore, a meta-analysis of antioxidant vitamin supplementation found that beta-carotene supplementation was associated with an increased all-cause mortality rate.[17] As a result of these European studies, as well as similar studies conducted here in the United States, articles in the *New England Journal of Medicine,* the *Journal of the National Cancer Institute,* and the *Lancet* all advise people to stop taking beta-carotene supplements.[18]

Taking extra vitamin A (retinyl palmitate and retinyl acetate) may be even more risky than using supplemental beta-carotene. Because beta-carotene is converted to vitamin A by the body, there is no reason a person eating a reasonably healthy diet should require any extra vitamin A. There is solid research revealing that supplemental vitamin A increases calcium loss in the urine, contributing to osteoporosis. One study found that subjects with a vitamin A intake in the range of 1.5

milligrams had double the hip fracture rate over those with an intake in the range of 0.5 milligrams.[19] For every 1 milligram increase in vitamin A consumption, hip fracture rate increased by 68 percent. Vitamin A supplementation has also been associated with a 16 percent increase in all-cause mortality.[20]

In spite of the huge volume of solid information documenting the deleterious effects of beta-carotene, vitamin A, and folic acid, it is still almost impossible to find a multiple vitamin that does not contain these substances.

Iron, Copper, and Selenium

Although iron is crucial for oxygen transport and other physiological processes, in excess iron is an oxidant that may contribute to cardiovascular disease and cognitive decline in older adults.[21] Iron should be taken as a supplement only when a deficiency or increased need exists, such as during heavy menstrual bleeding or pregnancy.

Recent studies have also shown that excess copper could be associated with reduced immune function, lower antioxidant status, atherosclerosis, and accelerated mental decline.[22] For these reasons, I also exclude copper from my supplements and those I recommend. A healthy diet gives us enough copper.

Selenium, of course, is essential, but a healthful diet gives us enough. There is some evidence that high selenium levels may contribute to diabetes, hyperlipidemia, prostate cancer, cardiovascular disease, and impaired immune and thyroid function. Therefore, supplementation beyond what is present in natural foods is likely not beneficial and may result in overexposure.[23]

Some people, even when consuming an ideal diet, may need more of certain nutrients. Individual absorption and utilization of nutrients varies from person to person, and some people simply require more to maximize their health. For example, B_{12} is always low on a near vegan or vegan diet, and some individuals require iodine or more vitamin D due to differences in absorption and utilization. Likewise, some people may also require more minerals such as zinc for maximizing their health and longevity.[24] I have designed a few supplements that conveniently supply the micronutrients of value, leaving out any potentially risky or controversial elements. These are available at www.drfuhrman.com.

I do recommend taking a high-quality multiple vitamin/mineral supplement to supply extra vitamins D and B_{12}, zinc, and iodine because ideal amounts of these are hard to acquire even in an excellent diet. As discussed above, the main issue here is making sure you do not ingest supplemental ingredients that are potentially harmful in your quest to optimize your levels of these valuable micronutrients. Most vitamins and minerals have a window of optimal intake, so the goal is to take the right amount, not too much or too little.

Vitamin D and Calcium

Even though most people are deficient in Vitamin D, too much can also be suboptimal, so blood tests are often recommended to assure the proper level of supplementation. The multivitamin/mineral I make available at www.drfuhrman.com contains 2,000 IU of vitamin D_3, an appropriate amount for the majority of individuals, even though some people could require more. Vitamin D is a critical nutrient, not just for your bones but also for general protection against heart disease, cancer, autoimmune disease, and many other health problems. Taking this higher level of Vitamin D has been shown to offer benefits to diabetics.[25] The ideal amount of vitamin D supplementation is best determined by blood work. The 25-hydroxy vitamin D level ideally should be between 30 and 50 ng/ml. I do not recommend high-dose calcium supplementation under the false pretense that it is good for the bones. In fact, too much calcium can weaken bones and can even contribute to calcifications in the vascular system.[26] Extra calcium is not required for people following my nutritional guidance because the diet has adequate calcium already.

Long-Chain Omega-3 (EPA and DHA)

I also recommend a supplement supplying the long-chain omega-3 fatty acids EPA and DHA. DHA is essential for optimal brain and eye function, and EPA is particularly protective and therapeutic for depression.[27] Deficiency poses increased risks to diabetics.[28] Conversely, adequate intake of long-chain omega-3 has been shown to benefit diabetics.[29] This can pose a challenge, however, because most DHA and EPA are derived from fish oil, and fish has been demonstrated to actually worsen diabetes.[30] In fact, fish intake showed a 22 percent higher risk for diabetes when comparing five or more servings per week with less than one serving per week.[31] Therefore, I recommend fish only in limited quantities, which are sufficient to assure adequacy in these fats. Though the body can make some EPA and DHA from walnuts, flaxseeds, and greens, the conversion is variable from person to person and the levels in most vegans are suboptimal. Besides, few people consume enough walnuts and flaxseeds every day to assure adequate DHA production. Supplemental DHA and EPA is a good idea for nutritional assurance, especially for diabetics.

However, I do not recommend fish oil capsules regularly because each capsule contains about 1,000 milligrams of oil, and this high dose of fish fat may have a negative effect on diabetes. The goal is to prevent deficiency, not supply excess. Instead, 150 to 300 milligrams a day is sufficient. Consider the algae-sourced DHA supplements available or DHA/EPA Purity, which provides a clean and effective source of these omega-3s without harvesting fish and without overdosing on fatty acids. I do not recommend the pharmacologic use of high-dose fish oil to lower cholesterol and triglycerides, because those dosages are not without risks.

Thiamine (Vitamin B₁)

Another supplement to consider taking if you have diabetes is thiamine (Vitamin B_1). Studies indicate thiamine deficiency often accompanies diabetes.[32] The higher the glucose levels, the more likely you are thiamine deficient, as thiamine is lost as glucose is excreted. Furthermore, even a mild deficiency of thiamine can promote complications of diabetes.[33]

Supplemental thiamine can help protect the kidney in the diabetic. Thiamine deficiency in diabetics is linked to increased oxidative stress and damage to kidney and nerves.[34] Studies have demonstrated that diabetics benefit from taking extra thiamine.[35] Diabetic nephropathy, neuropathy, and possibly retinopathy have been shown to improve with thiamine supplementation.[36]

I typically recommend that people with active diabetes in the process of reversal ingest about 10 to 20 milligrams a day of extra thiamine, that is about ten times the normal recommended daily intake. However, once the diabetic issues are resolved, this need not continue. Many health professionals use a much higher dose, which would only make sense if the glucose was very high and uncontrolled even with medications, but this is not the case with people following my nutritional guidelines.

Alpha Lipoic Acid

Even though alpha lipoic acid is commonly touted as an important supplement for diabetics, its use is questionable. When given intravenously, it has been shown to have some benefit to diabetic neuropathy, but its usefulness orally is not proven. Alpha lipoic acid is already available among the hundreds of beneficial compounds in green vegetables. The bottom line is that if you are following this program to reverse diabetes, it is unlikely that adding an alpha lipoic acid supplement will offer additional protection. However, people with neurological problems from their diabetes, especially if they are eating a diabetes-promoting diet, would be reasonable candidates for this type of therapy.

Glucose-Lowering Plants and Herbs

Supplementing your diet with herbs and plants to lower glucose is reasonable, both in capsule form and added into dishes. A few grams of cinnamon have demonstrated benefits at lowering blood glucose and improving lipid parameters without apparent side effects.[37] Gymnema sylvestre is also mildly effective.[38] Powdered fenugreek seeds require too large a dose to be effective so are more difficult to utilize. However, fenugreek extract has been found to lower fasting blood glucose and improve HbA1C in diabetics.[39] There are other plants and plant extracts that have little side effects and have been shown to mildly lower blood sugar. They include white mulberry leaf, banaba leaf, green tea, acacia extract, hops, bitter melon, and nopales cactus.[40]

Though these plant extracts are not as strong as blood-glucose-lowering medications are, the advantage of using natural agents is that, by themselves, they do not cause hypoglycemia. On the other hand, they do not address the main cause of diabetes. I find that most of my patients do not need these aids to adequately control their blood sugar because the diet and exercise program is so effective. However, in cases where extra help is needed, a combination of natural plant extracts can be added to the protocol, which can further reduce or eliminate the need for medications and their significant side-effect profile.

Plant Sterols and Pomegranate Extracts

Plant sterols and pomegranate extracts may also be considered for their lipid-lowering and cardio-protective properties in diabetics. Plant sterols are naturally present in plant foods (especially nuts and soybeans), are structurally similar to cholesterol, and are components of plant cell membranes, similar to cholesterol in animal cell membranes. Plant sterols have long been recognized, and are FDA approved, for their capacity to reduce LDL cholesterol.[41] More than forty human studies have been published confirming their LDL-lowering properties. Plant sterol supplements can produce a decrease of approximately 15 percent in LDL levels.[42] This LDL lowering occurs in the digestive system, where plant sterols inhibit cholesterol absorption. This blocks not only absorption of dietary cholesterol but also reabsorption of the cholesterol produced by the body.[43] An interesting recent finding is that plant sterols have additionally been shown to offer protection against several cancers.[44]

Pomegranates are a delicious and unique fruit that contain a wealth of beneficial phytochemicals. Their potent antioxidative compounds have been shown in medical studies to reverse atherosclerosis and lower cholesterol and blood pressure.[45] Among the antioxidant substances in pomegranates are anthocyanins, catechins, quercetins, and distinctive ellagitannins called punicalagins—punicalagins make up the bulk of the pomegranate's antioxidant load.[46] These potent antioxidative compounds are believed to be responsible for the pomegranate's numerous health benefits. In diabetics and nondiabetics alike, pomegranate reduces cholesterol, oxidative stress, and inflammation. In addition, pomegranate acts similarly to ACE-inhibiting drugs, naturally lowering blood pressure.[47] In one study of patients with severe carotid artery blockages, after taking one ounce of pomegranate juice daily for one year, on average these patients experienced a 12 percent reduction in blood pressure and a 30 percent reduction in atherosclerotic plaque. In striking contrast, the participants who did not take the pomegranate juice experienced a 9 percent increase in atherosclerotic plaque.[48] Also look for pomegranates in season, and if you can, freeze some of the seeds for use later in the year.

Chromium

Chromium is another nutrient commonly recommended to diabetics because those with diabetes are typically overweight and have been eating diabetic-promoting diets low in chromium. In other words, eating refined grains, sweets, and processed foods leads to chromium deficiency and worsens diabetes. A meta-analysis identified forty-one trials that evaluated the effects of various chromium formulations and dosages and found mild benefits, especially among patients whose diabetes was poorly controlled.[49]

Of course, when you eat a nutritarian diet with tomatoes, onions, and greens—foods that are very high in chromium—you protect yourself from chromium deficiency and diabetes. So even though clinical trials show a modest improvement in markers of insulin resistance and glucose levels in patients who supplement with chromium, I am not certain this supplement will be of value for people following my high-micronutrient approach, as they are now receiving adequate chromium and have already curtailed foods that cause chromium deficiency.

Remember, this is not a calorie-counting plan. You can eat as much as you desire of the recommended unlimited foods. You can eat limited amounts of several other foods too. Let hunger be your guide. Do not eat until you're full. Just eat so your hunger is satisfied. You do not have to eat all the foods in the suggested guidelines and menus, and you can switch around the meals if that's what works best for you. The most favorable way to eat is to learn from experience so you know the right amount of food that will allow you to be hungry in time for the next meal. If you are not hungry before the next meal, then you have eaten too much at the prior meal. Remember, hunger is felt mostly in the throat and increases your ability to enjoy food. If you are not hungry, do not eat. You will not die if you skip a meal or two or three. Err on the side of undereating so you can make sure you are eating only when you're hungry. You will enjoy eating more when you feel you have emptied your tank before refueling. As they say, hunger is the best sauce.

CHAPTER NINE

The Six Steps to Achieving Our Health Goals

I was diagnosed late November 2006 with diabetes at the age of fifty-two. My fasting blood sugar level at the time was 160. My weight was 218 pounds. I had been prediabetic for some time, along with a strong family history of the disease.

I took this news as a personal death wish. Being a pharmacist, I know the long-term complications and risks that diabetes carries. I became determined not to let this diagnosis doom me to a life of medication and routine insulin shots.

My wife found Dr. Fuhrman's book when she was browsing through the health section at our local bookstore. It was a godsend. While I have never been a big red meat eater, the thought of evolving my diet from a chicken and fish diet to the ultimate goal of a primarily plant-based diet seemed like a huge challenge. However, I was determined to beat diabetes at any cost.

I am proud to say that Dr. Fuhrman saved my life with his eating program. And I enjoy it. My latest test results speak for themselves: Cholesterol (total) 139, LDL 79, HDL 49; blood pressure 110/75; weight 172; A1C 5.3.

The only downside (if you call it that) to this experience was my need to buy a new wardrobe. My waist size dropped from thirty-eight inches to thirty-three inches. My large shirts now hang on me like a trash bag! I have lost the fat face that you see in my before photo. Thank you so much for giving me the means to change my life and become healthy. I am a new man!

—Steve D.

You have now read lots of science and logic. Hopefully, I have motivated you to make a change. The problem for some of us is that our behavior is not controlled by logic or science; it is controlled by feelings and emotions.

Change can be very difficult for some people and easier for others. Knowledge is clearly the key that can set us free from the vicious cycle of food addiction and self-destructive eating practices. If you are one of those people who finds change difficult, then let's work together to lose the emotional baggage that could be interfering with achieving excellent health. You have to lose the fear of change and the fear of giving up the comfort foods that fuel your food addictions. Replace them with excitement about a more pleasurable and comfortable life driven by the results you will see as your body is transformed.

We must look to the big picture and focus on long-term results, not short-term recreational eating. Together, we can get rid of the on-a-diet mentality and move to becoming an expert in nutritional excellence. Part of gaining this expertise is learning some great-tasting recipes that you will love. You also have to accept that it takes time to allow your taste buds to adapt to this new way of eating. Be patient with yourself. It could take a few months to prefer this way of eating and for your taste buds to sensitize themselves to lower levels of salt, sugar, and oil in natural foods. Over time, though, you will find that you enjoy this style of eating every bit as much, and even more, than your prior diet. This becomes especially true when you see the physical results of eating food that is so good for you.

It is a good idea to have a meeting with your family members or other people you live with to explain your new diet direction and to ask for their support. Do not try to change their eating habits, but ask them if they can help you by understanding what you are doing and why. Perhaps they can read this book to help you stay focused on your new diet style and new health goals. If your loved ones feel that you aren't trying to change them but are just asking them to respect, understand, and support you in your new diet plan, they may choose to make important changes toward better health for themselves.

Support is important. If you don't have supportive family or friends, or a supportive local network, join my website membership at www.drfuhrman.com to help you interact with others who eat the same way. The connection with others is valuable. You will find you are not alone. Thousands of others are on the same path to good health and want to support you and share your life's challenges.

Lastly, please don't forget the personal rewards and sense of achievement you will receive from earning back your excellent health. This is a reward that continues year after year, a reward that multiplies over and over again when other people are positively influenced to earn back their health because of your example. You do not live alone on an island. Your health and well-being, and the positive message you radiate to those around you, affect your family, friends, fellow workers, and community, even your nation. Destructive eating may give you a momentary high, but the most pleasure in life comes from more meaningful achievements. Make the commitment to earn back excellent health, and soon you will find that healthy eating can be just as pleasurable. Every day on this program places you closer to improved health and a more rewarding and pleasurable life.

Six Essential Steps to Help You
Reach Your Health Goals

Step #1: Make the Commitment and Write It Down

Keep track of your progress in a notebook. Start by writing down at least five reasons why you eat the foods you now do. Think of reasons why you should continue on your current course that has made you diabetic and is keeping you diabetic. This may sound like a silly exercise, but it is helpful for closure and clarity as you examine how you got to this point and how you plan to move toward health.

Then on the next page, write down the advantages of making a complete, firm, and irrevocable commitment to eating the way I am suggesting for the next twelve weeks. This means if you have diabetes, stay with the phase one dietary guidelines for the full twelve weeks. Allow time for your taste preferences to change, for your weight to drop, and for you to reduce serious health risks. Then, after the first twelve weeks, you can choose if it is the right time to make some small modifications to the program (moving to phase two) or to stay with phase one to maximize more gains. If the advantages are not immediately clear, take a moment and reflect on how your dietary habits are affecting you and everyone around you. What are the long-term dangers and consequences? This section will include why what you were eating was slowly killing you and what you hope to gain from this great change in food preferences and eating style.

Some people prefer to rationalize and protect their addictions. The desire to maintain these addictions has taken over their life despite the long-term health consequences. My guess is that these individuals did not get very far into this book. But the fact that you are still reading tells me you are ready. This step is very important because the rest of society eating the SAD will surely attempt to convince you otherwise. This notebook will prepare you for any internal conversations with yourself or any external conversations that may arise about your choices. It is the ultimate shield that protects you and enables you to confidently move toward health and a life that is not ruled by disease, medications, doctor visits, and fear.

The reason I chose twelve weeks for this firm commitment is because it takes us that long to change our taste preferences, learn new favorite recipes, and get in a rhythm of eating and enjoying this plan. Twelve weeks gets us past the stage of experimentation and into the realm of real everyday living. After twelve weeks, the health turnaround will be in full swing, and you will be seeing powerful results daily. Then it will be easy for you to feel and understand that this is a lifetime commitment to excellent health because this becomes the way we all prefer to eat and enjoy most.

After you have listed the advantages and disadvantages of your new diet style in your notebook, write down your commitment:

> I, (your name), commit to a new effort of health excellence over the next twelve weeks of my life. This includes eating only the healthiest foods, learning more about how to prepare them properly, and continuing without excuses or distractions consistently for the twelve-week period. I will also exercise every day.

A commitment means that an excuse is not an excuse. There can be no excuses to break this commitment. Addicts always have excuses. They always have a reason or rationalization to explain why their lifestyle change did not work or was too difficult.

"It is too difficult a time right now"; "I had to travel on my job"; "I am invited to a wedding"; "My son smashed up the car"; "I am having trouble at work" are all common excuses. The reality, however, is that life is stressful and change is difficult, but a commitment means you do it no matter what. Do not *try* to do this because trying means you will accept an excuse not to do it. Instead, commit no matter what life throws your way and no matter how much effort it takes. Things that have huge value require effort. Giving it a *try* is the formula for failure. Moderation means failure. Great success means a significant effort is usually required. Doing it *no matter what* is the formula for success. It is the formula that all successful individuals use to achieve great results.

Step #2: Draw Up a "Business Plan"

Regardless of the scale of any venture, a business plan is an invaluable asset. It is a map of how you plan to achieve your goals. I believe a plan is essential for success. Let's start by creating a weekly schedule. First work on the plan on scrap paper, and then write it in your notebook. The plan should consist of determining your food shopping days, what to buy and where, which days to do your cooking, what you cook, and what you plan to save as leftovers for the days you are not cooking. It should also include what days you exercise and what exercises you do on those days. All of these details need be part of your weekly planner. The more precise you are with the plan, the easier the follow-through as your days unfold. This step is essential for creating the important balance between your job, your diet, your exercise, and other details in your life. This plan will make sure this program is accessible no matter how busy you are.

Most people use weekends to do a big food shopping trip as well as their major weekly cooking. It's a good idea to make a big pot of veggie-bean soup, a salad dressing, and a cooked vegetable. Then you have food to use as leftovers for the main part of the week. You may also consider doing another small shopping trip and a food preparation session on a smaller scale, perhaps on an evening you take off from going to the gym. Sometimes frozen vegetables or canned beans can be used so you can exercise instead of cooking one night. When writing your business plan, choose which recipes you are going to make and the foods you have to buy to prepare them. Now you have your diet plan for the week.

Write your business plan in the form of a weekly calendar so you can see when you shop, cook, exercise, and do other errands and recreational activities. Planning is essential for success. You have to plan to make sure this program fits into your life. You don't always have to cook, but you always have to plan.

Step #3: Track Your Progress

Next, make a section in your notebook to track your medications, your weight, your blood pressure, your blood sugar, and your lab tests. Add an entry at least twice weekly documenting your progress, and include how many pounds away you are from your goal weight. Also write down the exercises you do and the length and vigor of the exercise so you can record your increases in exercise tolerance. This all will make it easy to reduce your medications at the appropriate rate and time.

It has been shown in medical studies that keeping a diet diary to record everything you eat helps with staying on a diet. This is not necessary for everyone, but I believe it is valuable to have the stats and details at your fingertips. Tracking progress can be a powerful motivator. Not only are you feeling better, more energetic, and losing weight, but also you have all the data to prove it.

Step #4: Make It Public

Now make your commitment public. Tell at least six people that you are making a radical change in your diet and your health. Tell them about this book, why you are making this commitment, and what you hope to achieve. Putting your commitment out in public makes it harder to turn back. Voicing your commitment to others makes it more established in your mind and makes it easier to resist the temptation of just eating your old way. It is common to receive a skeptical and discouraging response in return, sometimes even from your physician. That should only harden your desire and ability to prove them wrong. You are now in control of your life and your health. It is not up to your physician, friends, or family. Doing what everyone else does got you into this mess to begin with. Now it is up to you. You now know more than they do about getting healthy. You can do whatever it takes to earn back your health. Thousands of people have succeeded, and so can you. There is no turning back.

Step #5: Make Your Kitchen Healthy

Start by ridding your home of all unhealthy foods that you no longer intend to eat. Stocking your refrigerator and cupboards with the right ingredients for your diet and discarding the diabetes unfavorable foods is a major step. If you are living with people who cannot go along with excellent nutrition, separate the storage areas for your food. Use a different refrigerator for the unhealthy food or use a separate area of the main refrigerator. Get rid of or separate other food items as well. Put a big label on that section that says your name and *Don't Touch*. Give your spouse or others you live with permission to offer you help if you need it. They have to be instructed, "Don't let me eat the food that you are saving for yourself." Don't forget that if you are having problems committing to this program, it helps to talk with others who have done it and are doing it. You can find a support group on my website, www .drfuhrman.com.

Stock up on plenty of fresh and frozen fruits and vegetables. Buy many varieties of dried beans. Also buy some canned beans. Look for brands without salt, usually available in health food stores. The kitchen should become health central. It is vital to have it stocked and ready to help you help yourself. Roll up your sleeves, and let's get to work.

Step #6: The Exercise Prescription

It cannot be overemphasized that if you have diabetes, exercise is your prescription of choice. In place of dependency-inducing drugs, the proper medical intervention for this disease is to focus on the aggressive use of diet and exercise.

I always ask new type 2 diabetic patients, "How many days a week do you forget to take your medication?" They most often say, "Never."

Then I ask, "Why don't you just take it sometimes?" They look at me like I am crazy. Then I say, "Which do you think is more important to your long-term health, taking your medication daily or exercising daily? The answer is exercise—it is much more effective and more protective of your future health and survival than the medication. If you want to neglect yourself or forget to care for yourself, then forget the prescription drugs, but never forget to exercise." Too many people suffering from diabetic conditions believe that drugs are their savior. In reality, drugs can discourage us from taking the right steps toward good health, and the dependency on medication can be a downfall, not a savior.

Diabetes is a disease whose inherent causation is too little exercise and too much fattening food. The two key goals for anyone with type 2 diabetes are to get slim and fit. So why are more and more of us getting sick each year? As we discussed earlier, too many people are addicted to unhealthy foods and vicious toxic eating cycles. And too many are told by well-meaning physicians that diabetes is safely and effectively managed with drugs, not diet and exercise. Diabetics typically think (falsely) that their medication is life saving. They wouldn't dare miss it. The truth is that doing daily exercise is the real life-saving prescription. I make this clear to my patients and emphasize, "From now on, *never* miss your exercise. That is so much more important. It is critical to your recovery. You must be physically fit if you are going to beat diabetes."

In fact, according to large studies, diabetics who become fit can lower their risk of premature death by 40 to 60 percent depending on their body weight. One study, reported at the 2008 European Society of Cardiology Congress, showed that those diabetics who were highly fit had a 65 percent reduced risk of death in the seven years following the study compared to those with a low level of fitness.[1] Performing daily exercise and building up exercise tolerance are the most effective ways to enhance survival—their results are not matched by any medications to any degree.

There is no excuse for not exercising. Time is not an excuse. If you have time to take a shower, brush your teeth, and go to the bathroom daily, you can put aside ten minutes twice a day to exercise. Poor fitness is not an excuse. Even bodily injury rarely involves the entire body, so you can usually do some kind of physical activity. Even people in wheelchairs can exercise. And if you have poor exercise tolerance, that is even more reason to start.

If your blood sugar is running high, get it down quickly; address this right away with what goes in your mouth and how often you exercise. You must de-emphasize the role of medications and address your condition head-on. If you did not have medications, if they had never been developed, what would you do to bring your glucose down? You would exercise more and eat less, a much safer and more effective option than medication.

How to Exercise When You Have Diabetes

Okay, you are committed to getting healthy and fit to get rid of your diabetes. You are going to eat the right foods and exercise every day because you have finally decided to beat this disease once and for all. Now you know you should eat only when you're hungry and not eat so much that you are not hungry again for the next meal. My basic exercise rules for my diabetic patients work hand in hand with the understanding of true hunger. Generally if you eat three times a day, you should exercise three times a day. If you eat twice a day, you should exercise twice a day. Eat only when you are hungry, and that usually means eat only after you have exercised to work up an appetite.

Ideally, calories should be expended via exercise or physical activity in between meals so that before food is eaten, you have earned it. Exercising two to three times a day is usually necessary to achieve true hunger before a meal. The point is that you should not be eating food unless hunger demands you do. And then when you see how much better food tastes when you are truly hungry, you can appreciate that eating less and exercising more increases the pleasure of the food you do eat.

A good place to start your exercise regimen is with walking. The ultimate goal is working up to thirty minutes three times a day. Of course if you have not been walking regularly, we don't start out at that level. I recommend beginning with ten minutes three times a day. These short intervals make the exercise very easy to fit into our busy lives, and they allow us to quickly build up stamina over a few short weeks. If ten minutes is too easy, extend to fifteen minutes three times a day.

TEN EASY EXERCISES YOU CAN DO ANYWHERE

1. Walk briskly.

2. Put on some music and dance with a bouncing motion, transferring your weight from leg to leg.

3. Make-believe jump rope—jump in place as if holding a jump rope.

4. Get up and down from your chair 50 to 100 times.

5. Walk up and down a flight of stairs (or much more than one flight).

6. Do jumping jacks.

7. Hop around the room in a circle or back and forth in a line first on one foot, then on the other. Start out with 30 seconds per foot.

8. Rise up and down on your toes.

9. Stand on one leg and hold on to a chair or a wall for balance. Extend your free leg in front of you so the heel stretches out about 12 inches in front of your standing leg. Now, bend your standing leg knee so you lower your body about 6 inches, and then come back up. Do this 25 times and then switch legs. Repeat X times on each leg, depending on your fitness level and exercise tolerance.

10. Jog in place. Pick your knees up higher as you get in better physical condition.

Modify your exercise prescription to your individual capacity abilities and needs. Jumping is more vigorous than walking, so start out with only one minute of jumping or hopping if this is new for you. Also use a variety of the above exercise techniques (and many others) at each exercise session so that you involve a variety of skills and muscles. Start slowly, but do as much as you can handle comfortably.

The worse your physical condition and exercise tolerance, the more frequently you need to exercise.

If you are overweight and poorly conditioned, fatigue and soreness from exercise can be a limiting factor. The objective is to work up your exercise tolerance gradually. Walk, do a few flights of stairs, and then if you can't do anymore, wait a few hours and try again. The more out of shape you are, the more trouble you'll have doing much exercise, so the more frequently you'll have to exercise. If you can't exercise much at one time, you have to engage in shorter periods of regular but more frequent exercise. If you can only do a little exercise, such as five minutes or less, then plan on doing something at least four times a day. Exercise in spurts throughout the day. As time goes by, you will be able to increase the intensity and duration of the exercises. When you can spend an hour or more in the gym exercising vigorously, you can exercise less frequently.

You can burn calories, lower your blood sugar, and melt away fat with a variety of calorie-burning activities and exercises. However, calorie-burning activities such as walking, stair-climbing, biking, swimming, and using the elliptical machine are not sufficient. Weight training to increase muscular strength is also important. So often, diabetics complain it is difficult for them to lose weight even if they cut back significantly on their food intake. The way to address this is by combining the right diet with an assortment of exercises, especially muscle-strengthening exercises. Invariably, people who complain that their metabolic rate is low and they have trouble losing weight no matter what they eat have weak muscles and are poorly conditioned. Increasing their strength by weight training and doing other weight-bearing exercises creates an increase in muscle density, which helps to metabolize more calories. This critical increase in muscle density will help normalize metabolism, and as a result will address the problem that's causing diabetes.

Walking up flights of stairs is the very best exercise. Walk up as many flights of stairs as you can each day, and keep track of the total number of flights you do. Walking twenty to thirty flights a day is an effective way to meet your fitness goal. Most of my patients have a health club in their home—that is, a stairway. Many even have a second stairway going down to the basement. I ask them to walk up and down the two flights ten times in the morning and ten times at night. It takes only ten minutes, but it really works.

I also encourage patients to join a real health club and use a variety of equipment that uses many body parts for maximum results. The more muscle groups that are exercised, the more metabolically active players you have on your team to help you meet your goals. It is definitely helpful to have access to an assortment of exercise equipment, such as elliptical machines, treadmills, stair steppers, recumbent bicycles, and numerous resistance machines. When you tire of one machine, you can move on to a new one.

Strength-building exercise should be done daily too. However, the muscle groups exercised should be rotated so the same muscles are not exercised two days in a row. For example, on Monday, do exercises to strengthen your chest, shoulders, and middle back (latissimus dorsi). On Tuesday, do abdominals, lower back, and thighs. Wednesday, do biceps, triceps, forearms, upper back (trapezius), and calves. Thursday, start with chest, shoulders, and middle back again.

Of course, this is done in coordination with the other walking, running, jumping, climbing, stairs, swimming, tennis, racket ball, incline treadmill, biking, or other calorie-burning activities so as not to work the same muscle groups heavily two days in a row. For example, avoid stair climbing, elliptical, or biking the day after doing thigh-strengthening exercises. However, walking, treadmill walking, jogging, swimming, continuous dancing, and rowing machine exercises can be done every day in addition to strength training because these exercises will not make your thighs too sore. Ideally, I recommend my diabetic patients walk at least a mile every morning, exercise for ten minutes or so before lunch, and then exercise vigorously with jumping and strength training in the late afternoon or early evening before dinner.

It is also helpful to minimizing sitting during the day. If you work at a desk, consider purchasing a draft table which has a work surface at a height convenient for standing. Or work part of the day with your laptop or papers on an elevated counter so you can stand. Nowadays, you can purchase computer stands that rise up so you can work standing. If you're talking on the phone, stand up and walk as you talk. If you sit all day, you will make this program more difficult. Sitting all day is unhealthy, even if you exercise regularly. If you work standing and then sit for a bit, then work standing again, you will be more alert and efficient on the job while you're also training your body to be more fit.

Is Exercise Essential for Success?

Exercise is extremely important, but if your ability to be active and to exercise is limited, do not despair. My menu plans will still enable you to lose weight. People who are unable to exercise just require a stricter diet. Some people have health conditions that preclude them from much exercising. However, an exercise prescription can be devised to fit your capabilities. Almost everyone can do something; even those who cannot walk can do arm, abdominal, and back exercises with light weights or use an arm cycle. You can listen to upbeat music and rhythmically bounce up and down for a full song. Even if your full body weight does not lift off the ground, see if you can do some mild bouncing and hopping as you are dancing. Try to keep dancing for a full five minutes or more.

Exercise will facilitate your weight loss and make you healthier. Vigorous exercise has a powerful effect on promoting longevity. If you have the will to adopt this plan and take good care of yourself, you will find the will to exercise. Start slow and gradually work up, so you do not injure yourself. But immediately begin to do more than you are doing now.

You now know the nutritional science behind diabetes and why drugs are not the solution. You understand the ins and outs of what you should eat, and you have the six critical steps for preparation and achieving your health goals. It is now time to get slim and fit to prevent or reverse diabetes for good.

CHAPTER TEN

For Doctors and Patients

When Ricardo Pacheco started this program sixteen years ago, he had a fasting blood sugar of 175 and weighed 256 pounds. His blood pressure was 155/85, and he was taking 20 milligrams of Accupril daily for blood pressure as well as 15 units of insulin and 500 milligrams of metformin twice daily. At the first visit, I cut his insulin to just 10 units that first night and then just one more dose the following night with 5 units. The insulin was stopped after the two-day taper. Two weeks later, he weighed 237, a drop of nineteen pounds. His fasting blood sugar was 115, and his blood pressure was 125/80. About a month after that, he weighed 221, a loss of thirty-five pounds in fifty-two days. He had a fasting blood sugar of 80, which allowed us to stop the metformin at that time. His blood pressure was 88/70, so I discontinued his Accupril, which actually could have been cut out sooner. Luckily, he was not fatigued or lightheaded. He could have fainted or injured his kidney from the unnecessary medication. Ten months after the first visit, Ricardo weighed 190, a loss of sixty-six pounds since starting the program, and his HbA1C was 5.3 with a total cholesterol of 134 and a blood pressure *of 112/76. He was on no medication. He has been doing well, medication-free for over fifteen years.**

As we have discussed, diabetes mellitus is a tremendous financial and health burden on an already overstressed health care system. Diabetes and its complications contribute to an estimated total (direct and indirect) cost of $174 billion in the United States on an annual basis, including $116 billion in medical expenditures and $58 billion in lost productivity.[1] In 2011, according to the National Institutes of Health, the prevalence was 25.8 million, or 8.3 percent of the population.[2] It complicates the issue that approximately two-thirds of the U.S. population is overweight and/or obese, increasing the possibility of exponential growth in diabetes due to the higher likelihood of insulin resistance among this population.[3]

It is generally assumed that a combination of failure to lose weight, poor glucose control, and poor management of other risk factors increase the complications and risk of diabetes. Currently, medical care for type 2 diabetes consists of attempts to lower risk and achieve better metabolic control. Successful treatment outcome, however, is not consistently achieved with current drug-based recommendations.

In the 2009 consensus statement of the ADA and the European Association for the Study of Diabetes, the organizations recommend starting a nascent diabetic patient on lifestyle changes plus metformin. According to the authors, for most individuals with type 2 diabetes, lifestyle interventions fail to achieve or maintain the metabolic goals either because of failure to lose weight, weight regain, progressive disease, or a combination of factors.[4] Only about 36 percent of type 2 diabetics have achieved the ADA's recommended goal of a HbA1C <7.0, which means about 64 percent are still not reaching even the basic

---

*This was over 15 years ago. I am now especially careful in watching blood pressure, particularly when patients are on ACE inhibitors and can drop their blood pressure too low and injure their kidneys if the medications are not reduced or stopped in time.

(suboptimal) therapeutic goal.[5] These percentages are worse than in 1988 through 1994, when 44 percent reached the ADA goal.

Also disappointing is the finding of two large clinical trials, each with over ten thousand patients, that intensive medication therapy to tightly control glucose to near normoglycemic levels may not be the most effective treatment approach.[6] One trial was halted when data showed an increase in all-cause mortality (257 vs. 203) and no benefit in cardiovascular complications.

The problem is that the modern diet is so diabetogenic that most patients with type 2 diabetes do not achieve target glycemic levels with traditional therapies, and these agents are also associated with weight gain and poor tolerability.[7] Insulin therapy or intensification of insulin therapy commonly results in weight gain. Weight gain associated with insulin therapy is believed to be primarily due to the anabolic effects of insulin, an increase in appetite, and the reduction of glucose excretion in the urine. This weight gain can be excessive, adversely affecting cardiovascular risk profile.[8] These less-than-satisfactory results create a quagmire for the medical community. Diabetic care without substantially motivating patients to eat more carefully and do more exercise is suboptimal.

Heart Disease and Diabetes Have an Unbreakable Bond

Diabetics develop atherosclerosis early in life. They develop it even before the diagnosis of diabetes is entertained. Atherosclerosis, or the buildup of cholesterol and plaque inside of the blood vessels, is a disease created by excess caloric consumption. We can't separate the discussion about diabetes and heart disease completely from weight loss. Heart disease and heart attacks were exceptionally rare occurrences in human history until the explosion of commercial food manufacturing and processed food exposure in the 1900s. The low-micronutrient diet people eat today contributes to atherosclerotic plaque deposition in two basic ways. First, low micronutrient consumption promotes excess calorie intake, and second, low micronutrient intake increases oxidative stress and inflammation in the body, which further promotes atherosclerosis.

It is well-established that atherosclerotic plaque development and the factors that contribute to the instability of plaque that promotes clot formation are linked to inflammation-prone tissue. From the initial phases of fatty streak formation to the evolution toward plaque instability and rupture, the SAD, which gets its majority of calories from low-micronutrient processed foods and animal products instead of vegetables, beans, fruits, seeds, and nuts is the cause of this disease process. Circulatory disease, the leading cause of death in the modern world, is a dietary-caused disease that is most effectively dealt with from a dietary standpoint.

The impact of low-micronutrient eating takes its toll, promoting an inflammatory cascade underlying most diseases that plague the modern world. It is this combination of excess calories, fat deposition, and inadequate phytonutrients that creates a nation of cardiovascular-diseased individuals. It is possible, but much less likely, for thin people to develop atherosclerotic heart disease when eating a disease-promoting diet, but even most of these people still have significant amounts of abdominal adiposity and visceral fat.

In addition to my twenty-plus years of experience in treating advanced cardiac patients and diabetics with aggressive nutrition, my main scientific contribution to this body of knowledge is the explanation that the same underlying buildup of free radicals, AGEs, and other toxic agents caused by inadequate micronutrient intake not only create disease and promote tissue damage and aging, but they also promote overeating behavior, food addiction, and food cravings. The underlying drive to overconsume calories is just too difficult to address while addictive symptoms drive overeating behavior. This physical need for more frequent and concentrated calories creates emotional and thought rationalizations that seek to justify bizarre and illogical eating behaviors, leading almost everyone to overconsume calories. When a micronutrient-deficient diet is consumed, we desire an excessive amount of calories just to feel normal. There is no longer the connection between satisfying hunger and a normal body weight. So becoming overweight is due not just to easily obtainable calories and sedentary jobs but also to unhealthful eating that leads to addictive food-consuming behavior, resulting in overly frequent eating and overeating. The result is that the vast majority of Americans become overweight, atherosclerotic, and—now more and more—diabetic.

Using medical and surgical interventions while the underlying nutritional, biochemical, and lifestyle factors that caused the problems continue to percolate is doomed to failure. Medical care is expensive and futile compared to nutritional interventions which are remarkably effective for:

- Lowering cholesterol and lipid risk markers

- Improving vascular remodeling to facilitate oxygenation and to relieve and resolve angina

- Losing weight and glucose intolerance, reversing the diabetic process

- Reducing inflammatory and clot-promoting tendencies without incurring a risk of bleeding

- Reducing the tendency toward arrhythmia, sudden cardiac death, heart attack, and stroke

- Reducing all-cause mortality in all patients with all medical conditions

It can't be reinforced enough that the goal is a low body-fat percentage, not a low dietary-fat percentage. The low body-fat percentage is best achieved by prescribed regular exercise and nutritional excellence, and bringing back the connection with true hunger so recreational eating and eating outside of the demands of true hunger can be reduced.

Keep in mind that lowering cholesterol and losing weight do not adequately explain this high-nutrient diet's protective effects against cardiovascular disease. This prescribed diet effectively lowers high-sensitive C-reactive protein. This protein found in the blood has been proven to increase the risk of heart disease. In addition, this powerful diet offers vital anti-inflammatory protection and other beneficial biochemical effects. Even though drugs may lower cholesterol, they cannot be expected to offer the protection against cardiovascular events that superior nutrition can. The aggressive use of cholesterol-lowering drugs does not prevent most heart attacks or strokes and does not decrease the risk of fatal strokes.[9] In clinical trials, a significant percentage of patients who are taking the best possible statin therapy still experience cardiovascular events, such as heart attacks, sudden cardiac death, and strokes. Lowering cholesterol with nutritional excellence, however, can be expected to offer radically more protection and disease reversal than drug therapy can, without the risk or expense of prescription medication. I have seen the results in patients for more than twenty years, and now finally we are beginning to see the research results catch up and support my experience.

The reward for treating patients in this manner is to see improvements and disappearance not only of diabetes but so many other medical conditions as well. Headaches resolve, asthma episodes often go away, fatigue and body aches improve, digestive issues resolve, and most importantly, atherosclerosis and chest pains resolve without invasive procedures or surgeries.

Heart Disease Case Studies

The interesting part of the results achieved with excellent nutrition on this nutritarian diet style is that many of these participants were already on so-called healthy diets and were worsening before they followed my nutritional protocols to reverse heart disease. Some were even worsening on vegetarian diets. The other notable achievement is that people on this regimen do not see just a small reversal of atherosclerosis with excellent nutrition; they get a dramatic amount of reversal.

Case #1: David

David was a sixty-year-old man who had read the 1980s runaway best-seller *Fit for Life* over ten years ago and was following its recommendations for a starch-based Mediterranean-type diet. David ate mostly vegetarian foods with brown rice, potatoes, and whole wheat, fruits and vegetables, chicken only a few times a month, fish once or twice a week, and some olive oil on salads. He began an exercise program in June of 2006 and was surprised to find that he had chest pain with exertion.

His weight had been stable at 158 pounds for years. A thallium stress test indicated significant coronary artery disease with an LDL cholesterol of 126. His CT angiogram done on June 30, 2006, showed near total obstructive disease in the proximal left anterior descending artery due to low-density plaque.

David began my careful dietary protocol for the reversal of heart disease and did not have an angioplasty or bypass as was suggested by the cardiologists. After following my nutritional advice for one month, the chest pains resolved. His weight dropped to 140 in the first eight weeks. One year later, a repeat of the CT angiogram showed the left anterior descending artery with a non-obstructive mixed-density plaque with a stenosis estimated at 50 to 70 percent. David's weight has remained around 138 to 140 since following my dietary suggestions. In August of 2008, the last evaluation of his coronary arteries was performed, showing normal cardiac blood flow and no evidence of heart disease.

Case #2: Stan

Stan was a middle-aged male who had been on the strict version of a popular low-fat diet program that included meat but was largely based on vegetables, grains, and fruits for over three years while his carotid artery disease continued to worsen. After the first year on this seemingly healthy diet, the results showed no or very slight improvement. Stan continued on this very strict program for two more years to improve his disease. But it got significantly worse! The radiologist said, "The lesion on the left side is stable. There is some early buildup on the right side that has worsened, and I got a nice picture of a lipid [fat] inclusion in the artery wall."

Stan was then referred to me by the dietician at the health center that was monitoring his progress. After twenty months on my program, he saw great results. The radiologist's comments this time were: "Borderline evidence for atherosclerotic burden." He was no longer talking about a lesion or early buildup. There was barely any sign of atherosclerosis.

In twenty months on my dietary protocol, Stan lost ten pounds and is now 157 pounds. He is now running two miles per day, whereas he was running four miles a day during the last two unsuccessful years on his former low-fat flexitarian diet. "I just changed the emphasis off of lots of whole grains and onto high-nutrient foods. I felt lots better, I dropped my abdominal fat girdle without effort, and I enjoyed it," he said.

Case #3: Susan

Sixty-six-year-old Susan had a history of occasional irregular heartbeats noted to worsen with the ingestion of caffeine, alcohol, and sometimes even heavy exercise. She changed her diet first to vegetarian and then to low-fat vegan. After a little more than one year on the very low-fat vegan diet, her arrhythmias worsened, and she then developed atrial fibrillation.

Unsatisfied with the results of following the extremely low-fat vegan diet, she changed her diet to my nutritional recommendation. Susan improved the nutritional quality of her diet and added back nuts and especially seeds. Within three months, her cardiac arrhythmias completely disappeared.

Case #4: Debra

Debra arrived in my office as a seventy-two-year-old female type 2 diabetic at five-foot-one and weighing 160 pounds. She had been on insulin for twenty years. She was using 30 units of Lantus insulin at bedtime and 5 units of Lispro insulin before each meal, with blood sugar running between 125 and 175. Because Debra was taking under 50 units of insulin a day and the dietary intervention can account for at least 30 units of insulin, I discontinued all the insulin at that first visit and started her on 250 milligrams of Glucophage (metformin) three times a day. As it turned out, she did not feel well using even the low dose of Glucophage, and since her glucose levels were running between 110 and 130 in the few days after her initial visit, we decided to manage the diabetes without medications. In this case, we essentially stopped 50 units of insulin per day, and her blood sugar was better controlled without it.

Debra reported at her two-week follow-up visit that she felt much better off of the insulin. Her appetite was no longer ravenous, and she felt immediately lighter and more comfortable with walking and exercise. She lost nine pounds in those first two weeks. Debra is now safely off all medication and showing no signs of diabetes.

My high-nutrient-density diet is designed to be diabetic favorable, to reduce body fat, and to promote the regression of atherosclerosis. It accomplishes these goals for multiple reasons:

- No refined carbohydrates, neither sugars nor starches
- Minimal grains (intact grains only), 1 serving daily
- Very high fiber (over 50 grams per day)
- High viscous fiber (flax, oats, beans)
- High percentage of resistant starch
- Moderate fat from seeds and nuts
- Very low saturated fat

- Zero trans-fatty acids

- Sufficient omega-3 fatty acids

- High phytochemicals and antioxidants

- Low glycemic load

- Very low sodium (less than 1,200 mg/day)

- Low caloric density per food volume

- Minimal animal products, 3 servings a week (few ounces maximum) or less

It is designed as a therapeutic intervention for diabetics who want the most effective dietary regimen for maximizing health protection. Because the results have been so impressive, patient compliance has been favorable.

Results are aided when diabetic patients make a firm decision to attack their medical issue with complete dedication and effort to a high-nutrient diet.

Managing Insulin Use for Type 1 Diabetes

Nutritional excellence is critically important for type 1 diabetics. The combination between the disease and the SAD or even the ADA recommended diet results in needless medical tragedy in all type 1 diabetics.

FOR DOCTORS ONLY (BUT NONPHYSICIANS CAN READ IT TOO)

After adequate discussion, let patients know how much more rewarding and exciting it is for you as a physician to watch people getting well and beating diabetes and their other medical issues. After all, did we get into medicine to watch people deteriorate, or did we do it to help people get better?

Make a goal with the patient to shoot for at the next visit. Make it absolutely clear to the patient that diet and exercise are now the main means of glucose control, not drugs. Without an explanation and understanding of the futility of the drug-only approach and the absolute necessity of using diet and exercise to protect against further damage, patients are not given the proper opportunity to protect themselves. High glucose readings can be treated with enhancements in exercise and dietary adjustments much in the same manner doctors use medications. Medication used in the interim period until sufficient weight is lost should be limited to those drugs that are not counterproductive to losing weight or to saving and restoring pancreatic function, or at least moving in this direction.

This protocol essentially rules out the use of sulfonylureas and insulin, except considering insulin in very small amounts when the pancreatic beta reserve is unusually depleted. When the proper eating style is combined with the proper exercise program, medications are rarely needed, and even then only in small amounts. We also want to discontinue or at least reduce medications that can cause hypoglycemia since caloric reduction and increased exercise can reduce glucose so dramatically. Glucophage (metformin) and Januvia (sitagliptin) do not cause hypoglycemia and are safe medical options when medications are needed. Byetta (exenatide) is an option when the glucose is running too high on metformin and something stronger is needed, as it does not cause weight gain.

At the first visit, when patients begin this protocol, insulin and sulfonylureas should be reduced by at least one-half and discontinued in the weeks that follow, if possible. A very low dose of 250 milligrams of Glucophage three times daily can help people who experienced indigestion from higher dosages in the past. If the dose is increased slowly, the side effects are minimized. Byetta—injections of 5 milligrams twice a day—can be used in place of insulin in the initial phase if the glucose levels are unfavorable. In most cases, the oral medications such as Glucophage, Prandin (repaglinide), and Januvia can be used because they do not induce hypoglycemia or weight gain.

MOST FAVORABLE MEDICATIONS FOR USE IN CONJUNCTION WITH DIETARY TREATMENT OF DIABETES

Less Likely to Cause Hypoglycemia or Weight Gain

Glucophage (metformin)	Januvia (sitaglitin)
Byetta (exenatide)	Prandin (repaglinide)
Starlix (nateglinide)	Precose (acarbose)
Glyset (miglitol)	

LEAST FAVORABLE MEDICATIONS FOR USE IN CONJUNCTION WITH DIETARY TREATMENT OF DIABETES

More Likely to Cause Hypoglycemia and Weight Gain

Insulin (various types); ultra-long-acting Lantus and Levemir cause less weight gain

Amaryl (glimpiride)	Diabenese (chlorpropamide)
Glucotrol (glipizide)	Diabeta, Glynase (glyburide)
Actos (pioglitazone)	Avandia (rosiglitazone)

The goal is to avoid having a hypoglycemic event in the first week of dietary change. The glucose readings in the first few days of dietary adjustment should run 125 to 175; do not attempt tight glucose control.

It is safer to allow the patient to run a little high than to risk a hypoglycemic event when they start a diet this aggressive. I always give the first-time diabetic patients my cell phone number and ask them to call me every day for the first three days after their visit. On the third call, I determine when the next call will be or if they can wait until their follow-up appointment in two weeks.

I also instruct the patients who are reducing their insulin dose to cut back their dose considerably each time they get even one reading below 120. I emphasize strongly that if they don't, the next reading may have them in a dangerous hypoglycemic episode. I write out exactly which insulin to cut back on and by how much, reducing both the long-acting (Lantus or Levemir) and the short-acting meal-time insulin (usually Lispro). I carefully watch their morning fasting readings for guidance on the reduction of the long-acting insulin dose.

My type 1 diabetic patients wind up using approximately half the insulin dosages they required before working with me. They obtain favorable glucose and HbA1C levels, and get rid of the swings in glucose that require varying dosages of insulin. They are able to stick to an insulin dose without monitoring nightmares and constant adjustments. They have fewer fluctuations in numbers and avoid hypoglycemia, as their insulin requirements are now physiological, not pathological.

The most physiological insulin regimen is to use four shots a day: one of the long-acting, twenty-four-hour insulins (such as Levemir or Lantus) before dinner or at bedtime, and one short-acting insulin immediately before each meal. This most accurately mimics what a normal, nondiabetic pancreas would supply. Long-acting and short-acting insulins cannot be combined in the same shot, thus four shots are required per day, not three.

The nighttime long-acting insulin dose is usually cut back by 40 percent at the first visit, and the pre-meal (quick-acting) insulin is reduced by 30 percent. Because regular insulin extends its action beyond mealtime needs and can lead to hypoglycemia, it is no longer recommended as the medication of choice—especially for my patients, whose mealtime insulin requirements are even shorter lived.

Reviewing the morning fasting and preprandial glucose levels will help the physician adjust the bedtime long-acting dose, and the two-hour postprandial glucose readings will help further adjust the mealtime quick-acting insulin dose.

For type 1 diabetics, adjust the long-acting dose so the morning and preprandial glucose readings range from 80 to 120, and adjust the preprandial insulin dose so the two-hour postprandial glucose readings hit in the 130 to 175 range.

The only way to safely achieve these results without hypoglycemic reactions is not by conventional carbohydrate counting but by stability in the diet and stability in the insulin dosage. For example, a sample dietary skeleton for a type 1 diabetic woman with a daily intake of 1,500 calories would be:

Breakfast	400 calories	Two fruits, oats and oat bran, almonds, ground flax, walnuts
Lunch	500 calories	Salad with nut/seed-based dressing, veggie/bean soup, one fruit
Dinner	500 calories	Salad with hummus/bean/salsa dip, steamed greens, veggie stew, tofu or flavored beans or fish, one fruit

The consistency is in the food choices, as the carbohydrates used and the overall GI of the meal is low and the fiber is very high. The secret to the excellent control is the use of greens, beans, seeds, and nuts all together in the meal at both lunch and dinner.

This hypothetical type 1 patient will now require only 3 to 5 units of insulin before each meal and 15 to 25 units of long-acting insulin at night. Whereas under the standard carbohydrate-counting ADA regimen, the average type 1 diabetic would take 6 to 9 units before each meal and 40 to 50 units at night.

With consistency in diet and medication, precise management of type 1 diabetes is possible without highs or lows. Patients are no longer at high risk for heart disease, stroke, or other tragic complications of the condition. They are no longer overusing insulin. They are no longer destined to be overweight. Their lipids come under control, and they get better glucose management, without the risk of being hypoglycemic. Their condition is managed more physiologically, and they feel better.

Diabetes During Pregnancy

Gestational diabetes is a pregnancy-related condition affecting over 5 percent of pregnancies in which women without previously diagnosed diabetes develop high blood glucose in the diabetic range.

Because the early stages of diabetes have no symptoms, gestational diabetes is most commonly diagnosed by screening with a glucose challenge test (GCT) or a three-hour glucose tolerance test (GTT). American women are subject to intensive screening to identify gestational diabetes. Their blood is checked for an elevated glucose early in pregnancy; and then, at twenty-four to twenty-eight weeks, they are given a 50 gram GCT to test their blood sugar under the high glucose stress. If that test is suggestive, the glucose is elevated above 130 one hour later, a more definitive GTT is offered. The GTT is a longer test conducted over three hours, with blood drawn for glucose each hour. Only about a third of women who have an abnormal GCT are found to have gestational diabetes with the GTT. However the higher the number is on the GCT, the more likely the GTT will be positive.

Of course if a woman was already diagnosed with diabetes before pregnancy, she would not need to be screened, but it is more important than ever for her to adopt a diabetic reversal diet and begin to reverse her diabetic condition immediately. In the United States, where we have the most obese and diabetic-prone population in the world, this extensive and elaborate screening regimen is set up by obstetricians because gestational diabetes is more prevalent compared to other areas of the world where the diet is not so excessive and unhealthy people do not carry so much diabetes-promoting body fat.

Women are more prone to gestational diabetes if they are overweight before becoming pregnant. Then the placenta-produced pregnancy hormones, in conjunction with the increasing body fat, make the body more and more insulin resistant. People prone to diabetes in general are those with a more limited beta cell reserve in the pancreas.

Gestational diabetes develops for reasons similar to type 2 diabetes development in later adult life. As the insulin needs of the body increase during pregnancy, the beta cells can't produce sufficient insulin to keep up anymore. In this case, the heightened needs of pregnancy emulate the insulin needs in an overweight person. So diabetes diagnosed during pregnancy is predictive of an increased risk of developing diabetes in subsequent pregnancies and later in life, if the standard (disease-causing) diet is continued. One long-term study followed a group of women diagnosed with gestational diabetes for thirty years and found that half developed type 2 diabetes after eight years and more than 70 percent had diabetes after twenty-eight years.[10]

Some women are at such a low risk of diabetes during pregnancy that they would not need the glucose challenge. For them, a fasting glucose test would be sufficient. These are typically women who are already eating a healthful diet and have no family history or genetic tendency toward diabetes. They were a normal weight before pregnancy and have had no abnormal weight gain with pregnancy.

Medications Are Not the Best Solution

Women who are identified with gestational diabetes are targeted for nutritional counseling, and if that fails, they are placed on oral medications or insulin to lower their glucose levels. Gestational diabetes poses an increased risk to both mother and child. The reason there is such a concern about heightened glucose during pregnancy is that elevated glucose increases the size of the baby and results in delivery complications, often leading to the need for C-section. It also increases the risk of preeclampsia—high blood pressure during pregnancy and excess amniotic fluid around the baby. Babies of diabetic mothers are also at higher risk of underdeveloped lungs and respiratory distress after birth.

However, giving medications to lower the glucose in gestational diabetics is not effective or sufficient to lower the risks. The excess body weight and poor nutritional input that precipitated this problem still remain and are contributors to this risk even if the glucose is lowered with medication. For example, overweight women have children with more birth defects, especially dangerous heart defects. Children born to mothers with gestational diabetes are also more prone to obesity and diabetes themselves. Hypoglycemia after birth and increased risk of neonatal jaundice are also a concern in these babies.

Women who know about excellent nutrition, eat healthfully, and maintain a healthy weight before pregnancy and continue to eat right through pregnancy do not need to worry about gestational diabetes. Its high prevalence in our culture speaks to how poor our American diet is. I do not generally require my slim, healthy patients following my healthful dietary guidelines to even partake in the GCT or GTT. Excellent health and physical fitness are important throughout life—before, during, and after pregnancy.

What to Do If You're Diagnosed with Gestational Diabetes

The conventional approach to treating gestational diabetes is inadequate. The dietary advice most typically offered is simply not sufficient to bring the glucose into the normal range without drugs, and this is a time to act quickly and not play around with suboptimal advice. Given the dietary ineptitude, too many treating physicians often prescribe medications during pregnancy. Later on, labor induction and a C-section with a greater risk of neonatal admissions are the typical outcome.

Pregnant women are highly motivated for successful outcomes. It is my experience that they will carefully comply with the prescribed dietary regimen and achieve excellent results. Ideal nutrition is a valuable blessing to give to pregnant women to enhance their health, the safety of their pregnancy, and the health of their children. The problem here is that gestational diabetes is often characterized by very strong insulin resistance, so even when doctors prescribe insulin, large dosages are needed. Therefore, when prescribing a nutritional approach for this condition, an aggressive nutritional protocol is indicated, though not utilized by conventional physicians and dieticians.

This aggressive antidiabetic protocol is important to reverse gestational diabetes quickly and easily without risky drugs, which can induce neonatal hypoglycemia and preterm delivery. Then, if any medication is still necessary, it can be almost always limited to metformin, which is classified as the lowest risk medication to be used in pregnancy.

Avoiding medications whenever possible is wise anyway. Who knows the subtle long-term effects of these medications on your unborn child? Will those medications increase the occurrence of cancer in our children sixty years down the road? We just don't know. Less medication is better medicine, and no medicine is the best medicine. For women at a high risk of diabetes, with diabetes, or significantly overweight during pregnancy, I recommend this aggressive antidiabetic food plan at least until the glucose numbers return to a safe range. This is the same phase one advice for all diabetics attempting to attack their condition with full artillery.

A Sample Phase One (Aggressive) Diabetic Reversal Diet for Newly Diagnosed Gestational Diabetes

Do not make choices about what to eat and what not to eat just yet. For now, allow me to make these decisions for you. In order to give this method a chance to see what it can do, you must do it as exactly as prescribed, without modifications.

This menu includes options—do not attempt to consume all the food or dishes suggested. Choose only one or two options at each meal. You can eat the same dish for more than one meal and even cook extra to use as leftovers for a few days. Please follow this plan until your blood sugar is relatively favorable. Then once your condition is in the safe, nondiabetic range, you can follow the general recommendations in the menu section of this book.

Breakfast

Use one of the below suggestions per meal. Please note that carbohydrates (even beans) are not eaten with breakfast because insulin resistance is highest in the morning.

- A green salad with lettuce, thinly sliced red onion, tomatoes, and roasted red peppers with a creamy hemp seed herbal dressing made from hemp milk, seeds, and a fruit-flavored vinegar. Or a roasted tomato-basil dressing made from tomato paste, soaked dried tomatoes, raw and roasted garlic, vinegar, roasted red pepper, chopped scallions, basil, cumin, and cinnamon. Half cup of berries.

- A roasted eggplant casserole made with sliced zucchini, mushrooms, onions, tomatoes, garlic, and spices such as cinnamon and cumin, sprinkled with lightly toasted and chopped pumpkin seeds. Half cup of berries.

Remember, this breakfast seems unusual because with gestational diabetes, insulin resistance is elevated in the morning; so the treatment is a light breakfast of low-glycemic plant foods for breakfast—no grains or fruit, except some berries.

Lunch

Select two of the following options per meal.

- A vegetable-bean soup or stew, served over or with a bowl of shredded lettuce and shredded raw spinach. The soup should be made with a low-salt tomato and celery juice base and lots of leafy greens, leeks, zucchini, and onions.

- Roasted tofu slices or one ounce sliced turkey chopped with avocado, dill, and roasted garlic wrapped in raw collard green leaves.

- Zucchini-cauliflower casserole baked with chopped onions and mung beans or other sprouts and sprinkled with nutritional yeast.

- The roasted eggplant and mushroom dish from the breakfast menu can be eaten here instead.

- Spicy beans or lentils (1 cup) served hot over a bed of finely shredded lettuce and cabbage.

Dinner

Include all three options below per meal.

- A steamed green vegetable dish made with steamed or water-sautéed mushrooms and onions. Steamed green vegetables (string beans, artichokes, or asparagus), crushed raw walnuts, and lightly toasted almond slivers.

- Raw vegetables such as raw broccoli, snow peas, cauliflower, kohlrabi, cucumbers, radishes, peppers, tomatoes, or celery served with a hummus or salsa dip. A sunflower-mushroom burger served with lettuce, tomato, raw onion, and tomato sauce with steamed cauliflower or spaghetti squash.

- One fresh fruit for dessert, or two kiwis or a box of berries.

This is also a version of the phase one diet plan to follow for a person who wants the maximum results immediately and for the patient whose glucose is elevated and needs to get in immediate control. In other words, any diabetic beginning this program can start right here with this aggressive phase and then after a few weeks move on to the rest of the menu and recipe options presented in chapter 12. This would especially be the place to start if you are on medications presently, yet still have a fasting glucose level above 150. Once your numbers are more favorable, and you are successfully reducing medication, then you can move on to more menu and recipe options. Be careful because you must reduce or eliminate medications to prevent hypoglycemia, so follow your readings and adjust accordingly.

MESSAGE TO PHYSICIANS:

More and more physicians are expressing their support and enthusiasm for the nutritarian approach to treating diabetes and other diseases, and they are supporting this health-promoting mission. I invite other physicians to come to my office and observe the results firsthand. Those who have taken me up on the offer have often commented that the experience was actually more fruitful than anything they did in their residency training. Continuing medical education credits are available to physicians coming to conferences for further training. I am always looking for physicians all over the country that we can refer patients to. America needs physicians who have gained experience with tapering medications and are supportive in managing diabetic patients with superior nutrition excellence and exercise. I encourage interested physicians to contact me and to join the American College of Lifestyle Medicine, a physician organization designed to support such physicians.

CHAPTER 11

Frequently Asked Questions

I understand that for a lot of readers, this plan is a radical shift from your current approach to food. I am sure you have a lot of questions. Based on my years of leading patients through this diet, I have compiled the most asked questions I hear from patients following this diet protocol. I urge you to be patient and diligent, and I promise that the results will be life-changing.

What if I do not like eating this way?

To eat healthfully takes practice and perseverance. What makes change possible is a strong desire and motivation, a willingness to sometimes be uncomfortable, and perseverance to keep working on it. The more you make healthy meals, and the more days you eat these foods, the more your brain will naturally prefer to eat that way. Your taste for healthy, nutrient-rich foods will develop. A new food has to be eaten about fifteen times to make it become a preferred food. The more days you eat healthfully, the more you will lose your addiction to unhealthy stimulating substances. With time, you will look forward to—and prefer—eating a diet that is more natural and wholesome.

People can always come up with excuses why something is too difficult to do, and your subconscious mind may promote this. But a strong desire and commitment to achieve your health and weight goals can silence these objections. With planning and support, you can overcome every obstacle. I promise that eventually you will prefer your new diet and newfound health.

There are over fifty great-tasting recipes in this book. Try them. There are hundreds more available on my website and in my other works. I promise that you will find many recipes you love.

Why is it so difficult for me to give up foods that I know are destructive to my health?

To be successful in achieving optimal health and permanent weight loss, we have to consider the complexity of human nature. We have discussed how a disease-promoting diet can be physically addicting, resulting in some mild discomfort during the first few weeks of the change. Of course, food is addicting physically and emotionally, so much so that some people would prefer death to diet change. The subconscious mind does not always care about logic and science. We are physical, emotional, and social beings. These factors must be considered. If not, many people will reject learning more about a health-giving lifestyle despite having an interest in gaining more information. Some will come up with rationalizations or excuses why they can't change.

It is not uncommon for people to give reasons and excuses to continue down the road of dietary suicide. This is a manifestation of a subconscious process. Our brains are designed to dim awareness to information that causes us distress and anxiety. For many people, the thought of changing the way they think about food and the way they eat is a source of anxiety in itself. Unhealthy foods are a slow-working poison, but our mind fears change even more than it fears this poison.

The addictive mind can always create a reason to justify the continuation of their addiction. For example, food addicts may think, *If you had a life like mine, you would binge too.* They get into a self-defeating cycle of self-pity and gloom. Once addictions take hold, they lose total control of their decision making, and continuing the addiction returns to the top of the subconscious agenda. Overeating, eating poorly, and remaining overweight in spite of health consequences are examples of the power food addiction has to control the subconscious mind.

Overeating is often the direct result of low self-esteem, which makes people vulnerable to negative peer pressure, addictions, and emotional binging. Some people may fear appearing different from others and believe that changing the way they eat will result in a loss of social status. Also, people sometimes overeat to raise dopamine and serotonin activity in the brain so that they can dull the frustration, disappointment, and pain of life.

Nevertheless, all the negativity can be turned around. Changing your diet can go hand in hand with a new attitude about yourself, life, and the possibilities in front of you. A healthy diet goes hand in hand with a healthy attitude about life and a love of life.

Acknowledge the conflict, acknowledge the difficulty, and just do it. You must face facts, accept your discomfort, and work through it. The subconscious mind might not like the changes you are making, but you have to hang in there until the change feels natural. You can't cure an addiction unless you are willing to fight through your internal debate and move on. Addiction thrives in isolation. The keys to getting over it are to make the commitment to stick with your new program and to acknowledge that commitment to people you know. It gets easier and more pleasurable every day that you move toward better health.

Will I feel unwell from withdrawal when I first start out? How long will any negative sensations last?

As your body's detoxification activities increase in the first week or two of this program, the symptoms of toxic hunger could increase. These feelings could include lightheadedness, fatigue, headaches, increased urination, sore throat, flatulence, and, very rarely, fever, body itching, or rashes. These uncomfortable sensations rarely last longer than one week, and even if they do, they will lessen with time.

Occasionally people find it takes time for their digestive tract to adjust to all the additional raw fiber. They experience an increase in gas or bloating or have looser stools. This is usually caused by the increased volume of raw vegetables and because you swallow more air when you eat salad than you would eating other foods. It is remedied by chewing better so the air is out of the food before swallowing it. Better chewing also breaks down the cells, making them easier to digest. If you encounter this problem, increase the amount of raw salad you eat only in gradual amounts. You can also eat fewer raw foods and more cooked vegetables, such as steamed zucchini, squash, artichokes, peas, and intact whole grains such as steel-cut oats and wild rice. When the symptoms subside, gradually increase the amount of raw greens and cruciferous vegetables in your diet.

Beans, nuts, and seeds can also contribute to digestive problems at the beginning of transitioning to this way of eating. To combat these issues, use beans at almost every meal but in very small amounts until you adjust. Make sure the nuts and seeds are spread out at various meals and you are not eating too much at one time.

If you are troubled by digestive problems, try the following:

• Chew your food better, especially salads.

• Eat beans in smaller amounts.

• Soak beans and legumes overnight before cooking.

- Avoid carbonated beverages.

- Do not overeat.

- Eat fewer raw vegetables and more cooked green vegetables, and then increase your raw vegetables gradually.

Be patient and give your body a little time to adjust to a different eating style. Remember, your digestive tract can make the adjustments if it's allowed to do it gradually.

Isn't diabetes mostly genetic? It isn't my fault—my whole family is overweight.

The role that genetics plays in obesity and diabetes is undeniable. People of certain descent have a smaller beta cell reserve in the pancreas. If these individuals eat an American-style diet and become overweight, they have a dramatically higher risk of becoming diabetic.

That said, the doubled rate of diabetes in this country in the last twenty-five years obviously did not occur because of genetics. Even though some people may be at a high genetic risk of developing diabetes, this does not mean that their health fates are predetermined. In fact, the reason for the increased genetic risk is likely because their ancestors were all thin and highly active. They did not require a large reserve of beta cells in the pancreas. This inheritance pattern is only a risk factor for people who eat a disease-promoting diet. Although people whose parents are obese have a tenfold increased risk of becoming obese themselves, the explosion of obesity is a recent phenomenon in human history brought on by fast food and the processed food industry. Clearly, obesity is not primarily genetic.

So it is the combination of food choices, inactivity, and genetics that determines obesity and diabetes. Those who genetically store fat more efficiently may have had a survival advantage thousands of years ago when food was scarce. But in today's modern food pantry, where high-calorie, toxic foods abound, those people are at a survival disadvantage. Focusing on the element of genetics in the formula doesn't solve the problem. You can't change your genes. Rather than taking an honest look at what causes diabetes, Americans are still looking for a magical, effortless cure for it—a gimmick, drug, or surgery. The only answer is living a healthy lifestyle focused on excellent nutrition along with adequate activity and exercise. When you live this way, the benefits will overwhelm genetics and allow even those people with a genetic predisposition to weight gain and diabetes to achieve a healthy weight and a long, disease-free life.

One of the most exciting studies in the field of weight control and obesity in recent years was published in the *New England Journal of Medicine*.[1] This study documented that if you have a friend who is obese, your risk of developing obesity increases by close to 60 percent, a higher rate than if a sibling or even a spouse becomes obese. This high percentage held up even after controlling for the fact that people tend to form bonds with others similar to themselves. If both people listed each other as friends, and one became obese first, the second was approximately three times as likely to follow suit.

This finding illustrates that obesity is spread by similar eating styles in social networks. Peer influence is not to be underestimated. However, understanding how powerful bad influences can be, especially with society's approval and promotion of addictive eating, leads to the inescapable conclusion that healthful behaviors can be just as contagious if you are surrounded by health-conscious people. One powerful secret to a slim body and good health is to cultivate friends who are supportive and can share a healthy eating style with you. Genetics are not the major factor. The social norms of the modern world have made obesity and diabetes pervasive.

I don't have time to cook. Do you have any tips for quick and easy meals?

You do not have to use fancy and complicated recipes all the time. Simple foods are quick and easy and can work in this program too. Consider some of these options to make your diet easier and more convenient.

Breakfast

Combine fresh fruit in season, or even frozen fruit, with raw nuts and seeds. You can soak some oatmeal in water overnight—no cooking is needed. Then add some berries and walnuts and a splash of unsweetened soy, almond, or hemp milk and you have a meal. You can make a quick smoothie with one fruit, a handful of greens, a half cup of unsweetened hemp milk, and flax seeds. It is okay to eat even lighter for breakfast and have just one or two fruits.

Lunch and Dinner

Your basic lunch or dinner should be a salad with a healthy dressing and a bowl of vegetable or bean soup. Make a quick salad from premixed and prewashed greens. Add chopped nuts and cut fresh fruit or no-salt canned beans, and top it with a low-fat and low-salt dressing, dip, fresh lemon, or balsamic or flavored vinegar. You can also eat raw vegetables and avocado with a low-salt salsa or hummus dip. I often heat up a little soup and pour it over some shredded lettuce as my lunch meal. I have also developed bottled dressings for this program, and they can be purchased and shipped anywhere.

Including some defrosted frozen vegetables or fresh or frozen fruit is a good way to round out a meal, salad, or leftover dish. Try steaming some fresh broccoli, spinach, or another green vegetable and adding a no-salt seasoning. Before cooking, apply a little olive oil to your hands and rub the raw vegetables, and then add chopped garlic and onion. A quick and tasty dish can be made from whatever fresh vegetables you have on hand. Use a no-salt canned soup and add your own steamed or frozen vegetables to increase the nutrient density.

Another easy meal option is to make a meal with a low-salt tomato sauce as the base. I often mix tomato sauce with some almond butter and a little gourmet vinegar, usually fig vinegar, and then use it as a dressing or dip for the vegetable dish, either cooked or raw.

When you do have the time, cook in large batches and save leftovers for other meals.

How can I stay on this diet style if I have to eat out in a restaurant?

Choose restaurants that have healthful options, and get to know the places that will cater to your needs. When possible, speak to the management or chef in advance. Eat early, before the restaurant gets very crowded, so the staff will have time to modify a dish or make something special for you. If you eat out for breakfast, avoid bread, bagels, and breakfast sweets. Go for oatmeal and fruit instead. For lunch and dinner, ask for a side of steamed vegetables instead of pasta or white rice to accompany your main dish.

In Asian restaurants, order vegetable dishes that are steamed with the dressing or sauce on the side. Because soups are made in advance in restaurants and are always very high in salt, it is best to stay away from them. Go with salads, simple vegetarian entrées, or on occasion plain broiled fish, shrimp, or chicken prepared without empty-calorie sauces. I often order a double-size salad and pay double for it. Tell the food server not to bring bread to the table to avoid the temptation to fill up before the meal. Always ask for the salad dressing (which is usually high in salt) on the side and use sparingly. Or better yet, use balsamic vinegar or lemon juice.

You can follow this diet on the road if you are committed to success. It just takes some diligence to plan where to go and to make sure in advance that you will have the foods you now eat. If you're staying at hotel, it is not difficult to eat healthy. When traveling, I go to a supermarket and purchase lettuce, frozen vegetables, fresh fruits, and canned beans to bring to my hotel room. I request a small refrigerator in my room, and I travel with a bowl, a can opener, silverware, and raw seeds and nuts in my suitcase. Then I can eat healthy canned beans, canned veggie-bean soup, frozen veggies, and frozen fruit tossed on a salad. I often use the electric coffee maker that is usually in the hotel room to heat up soups and oatmeal.

How much salt can I use safely? What about sea salt?

This book is designed for people who want to reverse their diabetes, lose weight, regain their health in general, and prevent further disease. If you are looking to get rid of diabetes and high blood pressure, any added salt—outside of what is contained in natural foods—is likely to hinder your recovery.

Sea salt is 99 percent chemically the same as table salt and is just as harmful. The very small amount of minerals found in special salts, touted for their health benefits, is insignificant and does not make consuming a high amount of sodium any less harmful. Salt consumption is linked to stomach cancer, high blood pressure, and heart attacks. For optimal health, I recommend that no salt be added to any food. The famous Dietary Approaches to Stop Hypertension (DASH) study clearly indicates that Americans consume five to ten times as much sodium as they need and that high sodium levels over the years have a predictable effect on raising blood pressure.[2]

Salt also pulls out calcium and other trace minerals, which is a contributory cause of osteoporosis.[3] If that is not enough, high sodium intake is predictive of increased death from heart attacks. In a large prospective trial published in the *Lancet,* there was a frighteningly high correlation between sodium intake and all-cause mortality in overweight men.[4] The researchers concluded, "High sodium intake predicted mortality and risk of coronary heart disease, independent of other cardiovascular risk factors, including high blood pressure. These results provide direct evidence of the harmful effects of high salt intake in the adult population."

There is controversy regarding the use of salt, as well as other points in human nutrition. Some studies indicate salt is not bad for health. However, those findings are always distorted by looking at a sickly population with high blood pressure and atherosclerosis that then attempts to lower salt intake later in life, when much of the damage is already done. It is always wise to err on the side of caution and not to incur needless risk. The earlier in life you begin the right choices, the greater the benefits. You would not want an error in judgment to cost you your good health or your life. For a hundred thousand years, our human ancestors and pre-human primates never salted their food. This is a recent phenomenon in the history of the human race, but natural plants supply all the sodium we need, as they have for those thousands of years. I heard gorillas go light on the salt shaker too.

My thousands of patients and website members have achieved dramatic reversals of heart disease, diabetes, and high blood pressure, and I am certain that removing salt from their diets was an important factor contributing to their health accomplishments. That said, a small amount is permissible; those parameters are discussed further below.

It's best not to add salt to foods, and I recommend looking for salt-free canned goods and soups. Because most salt comes from processed foods, bread, and canned goods, it shouldn't be hard to avoid added sodium. Use herbs, spices, lemon, vinegar, or other no-salt seasonings to flavor food. Condiments such as ketchup, mustard, soy sauce, teriyaki sauce, and relish are very high in sodium. If you can't resist them, use the low-sodium varieties sparingly.

Ideally, all your foods should have less than one milligram of sodium per calorie. Natural foods contain only about half a milligram of sodium per calorie. If a food has 100 calories per serving, yet contains

400 milligrams of sodium, it is a very high-salt food. If a 100-calorie serving has less than 100 milligrams of sodium, it is a food with little added salt and can be used in your diet. Overall, the guideline is for women to consume not more than 1,000 milligrams of sodium a day and men not more than 1,200. Since a natural food diet gives you 500 to 800 milligrams a day, that leaves you only a few hundred extra to be flexible with.

If you don't use salt, your taste buds will adjust with time and your sensitivity to taste salt improves. When you are using lots of salt in your diet, it weakens your taste for salt and makes you feel that food tastes bland unless it is heavily seasoned or spiced. The DASH study observed the same phenomenon that I have noted for years: it takes some time for a person's salt-saturated taste buds to get used to a low sodium level. If you follow my nutritional recommendations strictly, without compromise, avoiding all processed foods or highly salted foods, your ability to detect and enjoy the subtle flavors in fruits and vegetables will improve as well. Over time, you won't miss the salt and won't even want it.

What about alcohol—should I be drinking red wine for my heart?

Alcohol is not actually heart-healthy; it simply has anticlotting effects, much like aspirin. Researchers have found that even moderate consumption of alcohol, including wine, interferes with blood clotting and thereby reduces heart attacks in high-risk populations, such as people who eat the typical, disease-promoting, American diet. Thinning the blood with alcohol or aspirin is not health-enhancing unless you are eating the typical heart-attack-inducing diet. Once you are protected from heart attacks and strokes with superior nutrition, the blood thinning only adds risk in the form of gastrointestinal bleeding or a hemorrhagic stroke. Red wine contains some beneficial compounds such as flavonoids and resveratrol, a potent antioxidant in the skin of grapes associated with a number of health benefits. Of course, grapes, raisins, berries, and other plant foods also contain these beneficial compounds. You do not have to drink wine to gain these benefits.

Moderate drinking is defined as a maximum of one drink per day for women and two drinks for men. Consuming more than this is associated with increased fat around the waist and other significant health problems.[5] But more concerning is the link between drinking and cancer. Even a moderate amount of alcohol may also increase the risk of breast cancer.[6] A review of meta-analyses reported in 2009 concluded that one drink per day increased breast cancer risk between 7 and 10 percent.[7] More recently, a careful twenty-eight years of follow-up of the Nurses' Health Study found a significant increase in risk at even lower levels of consumption: the range of 5 to 9.9 grams per day (three to six drinks per week) was associated with a 15 percent increase in breast cancer risk.[8]

Another problem with alcohol consumption, especially at more than one drink a day, is it can create mild withdrawal sensations the next day. These sensations are commonly mistaken for hunger, which leads people to overeat. Because of this, moderate drinkers are usually overweight. Furthermore, recent studies have also shown that even moderate alcohol consumption is linked to a significantly increased incidence of atrial fibrillation, a condition that can lead to stroke.[9]

Overall, it is safer to eat a diet that will not permit heart disease. Don't rely on alcohol to decrease the potential of blood to clot. Strive to avoid the detrimental effects of alcohol and to protect yourself from heart disease with nutritional excellence. Having one or two alcoholic drinks a week or a few glass of wine per week is not a major risk, nor is it a major health asset. However, if consumed in excess, it can develop into a significant health issue.

Is it important to eat organically grown foods for good health?

The concern implicit in this question is about pesticides, and it is a real one. The Environmental Protection Agency has reported that the majority of pesticides now in use are probable or possible causes of cancer. Studies of farmworkers who work with pesticides suggest a link between pesticide use and brain cancer, Parkinson's disease, multiple myeloma, leukemia, lymphoma, and cancers of the stomach and prostate.[10] However, does the low level of pesticides remaining on our food present much of a risk? Some scientists argue that the extremely low level of pesticide residue on produce is insignificant and that there are naturally occurring toxins in all natural foods that are more significant. The large amount of studies performed on the typical pesticide-treated produce have demonstrated that consumption of produce, whether organic or not, is related to lower rates of cancer and increased disease protection. In short, it is better to eat fruits and vegetables grown and harvested using pesticides than not to eat them at all. The health benefits of eating phytochemical-rich produce greatly outweigh any risks that pesticide residues might pose.

It should be recognized that fruits and vegetables are not all subject to the same pesticide exposure. The following chart shows the pesticide breakdown by food in order of pesticide content. Spinach, strawberries, and celery have the highest pesticide residue and are the most important foods to consume organically grown.

If available, organic food is certainly your best bet to limit exposure to toxic chemicals. Even if you can eat organic versions of only the top twelve most contaminated fruits and vegetables, you can reduce your pesticide exposure by about 90 percent. In addition, organic foods usually have more nutrients than their conventional counterparts.[11] They also taste better and are generally better for farmers and the environment. Remember that cutting out processed foods, canned foods, and animal products already reduces your exposure to chemicals, toxic

Twelve Foods with the Lowest Pesticides

1. Onions
2. Sweet Corn
3. Pineapple
4. Avocado
5. Cabbage
6. Sweet Peas (frozen)
7. Asparagus
8. Mangos
9. Eggplant
10. Kiwi
11. Cantaloupe (domestic)
12. Sweet Potatoes

Twelve Foods with the Most Pesticides

1. Apples
2. Celery
3. Bell Peppers
4. Peaches
5. Strawberries
6. Nectarines (imported)
7. Grapes
8. Spinach
9. Lettuce
10. Cucumbers
11. Blueberries (domestic)
12. Potatoes

compounds, pesticides, antibiotics, and hormones, so that is a huge step in the right direction, even if your entire diet is not organic.

Why not olive oil? I thought it was good for me.

No oil should be considered a health food. All oil, including olive oil, is 100 percent fat and contains 120 calories per tablespoon. Oil is high in calories, low in nutrients, and contains no fiber. It is the perfect food to help you put on unwanted and unhealthful pounds and sabotage your plan to rid yourself of diabetes and extra body fat.

Oil is a processed food. When you chemically extract oil from a whole food (such as olives and various nuts and seeds) you lose the vast majority of nutrients and end up with a fragmented food that contains little more than empty calories. You do need some fat in your diet, but when you consume whole foods, such as walnuts, pistachio nuts, sesame seeds, chia seeds, and flaxseeds instead of their extracted oils, you get all of the fibers, flavonoids, several phenolic compounds, sterols, stanols, and other nutrients they contain as well as all of their positive health benefits.

Foods rich in monounsaturated fats like olive oil are less harmful than foods full of saturated fats and trans fats, but being less harmful does not make them healthful. The beneficial effects of the Mediterranean diet are not due to the consumption of olive oil; they are due to antioxidant-rich foods including vegetables, fruits, and beans. Pouring a lot of olive oil onto food means you're consuming a lot of fat. Eating a lot of any kind of oil means you're eating a lot of empty calories, which leads to excess weight, which leads to diabetes, high blood pressure, stroke, heart disease, and many forms of cancer.

When you eat oil instead of seeds and nuts as a source of fat, the fat is absorbed and enters the bloodstream rapidly. Because you cannot utilize the surge of calories, the body efficiently stores them as fat on your body within minutes. When you consume oil, the calories go from your lips to your hips within minutes, and as you know by now, fat stored on the body promotes disease and causes diabetes. When you instead consume some seeds and nuts as a source of fat, the fat is bound to sterols, stanols, and other fibers that slow the absorption of the fat over many hours and limit all the absorption of the fat contained, enabling the body to burn the fat for energy.

While nuts and seeds have anti-inflammatory effects, oil has pro-inflammatory effects, so it is a horse of a different color. For example, a study compared an olive oil–containing Mediterranean diet with one which substituted pistachio nuts as a source of fat. The researchers documented improvement in endothelial function (health of the lining of the blood vessels). They noted reduced inflammation, increased youthful elasticity of the blood vessels, lowering of the cholesterol and triglycerides, as well as a lowering of blood glucose as a result of eating the nuts in place of oil.[12] Plus, as discussed earlier, whereas oil promotes weight gain, when nuts or seeds are substituted for oils or carbs without increasing the overall caloric load, the result is lower glucose, lower cholesterol, and lower body weight.[13]

You can add a little bit of olive oil to your diet if you are thin and exercise a lot. However, the more oil you add, the more you are lowering the nutrient-per-calorie density of your diet—and that is not your objective, as it does not promote health.

Hopefully the answers in this chapter have proved useful as you begin your journey toward health. This approach may sound complicated at first, but I assure you that it will start to make sense very quickly. I have devised a very simple formula of nutrients per calorie. In the next section, I map out some meal plans that will allow you to see how to start building your own personal menu. Your health begins now.

CHAPTER TWELVE

Menus and Recipes

These sample meal plans provide 1,400 calories per day and are designed for moderate weight loss. You can follow these menus closely or just use them as guidelines. Feel free to substitute other healthy recipes from my recipe list in this book.

If you are a slim type 1 diabetic male and do not need to lose weight, you may add 1 to 2 additional ounces of raw nuts or seeds, some avocado, or even some additional starchy vegetable or grain such as wild rice, whole grain barley, quinoa, peas, corn, sweet potato, or squash. In other words, you can easily adjust these menus to your caloric needs by increasing some of the foods that generally would be more limited for overweight individuals looking to maintain a lower caloric level.

On the other hand, some people may require even fewer calories than are present in these menus. Remember, you can always eat less food. You should not eat if you're not hungry. When you eat so healthfully, you do not need as much food as you used to think you needed, and you will feel better and be satisfied with less. Try to eat lightly so you can feel the pleasure of getting hungry again in time for your next meal. If you consume too much with breakfast, you will not be hungry for lunch. If you consume too much food for lunch, you will not be hungry for dinner. Try to eat only enough to hold yourself over to the next meal, so you can get that full satisfaction from eating only when you're really hungry. You do not have to eat three meals a day. I consider a snack (breakfast) and two meals (lunch and dinner) enough for most people.

The foods in these menus are not weighed and measured. That is purposeful. It is necessary for you to live with and adjust this diet style under your own control. The size of the portions and how much you eat must be controlled by your body and mind, applying the knowledge you've learned here and achieving a stable plan that you can live with for years.

When you eat a diet with this degree of nutritional excellence, it enables you to be comfortable eating fewer calories. Hopefully you will have lost your desire to overeat and you have the knowledge you are healing your body as you do so. It is still possible for some people to overeat, due to emotional habits and other self-destructive thought processes. The most effective treatment breaking any bad habit or addiction is abstinence. In other words, this gets easier the longer you do it. I encourage you to err on the side of caution, don't overthink this, just do it, and give this new way of eating a chance to become your new habit and preferred way of eating.

Remember, a slower metabolic rate is not a bad thing. The slower your metabolism, the slower you are naturally aging. One of the mechanisms through which eating less prolongs life span is by slowing the metabolic rate. The slower your metabolism and the lower your body temperature, the slower you are aging. So rid your mind of the myth that you have to eat more or more frequently to raise your metabolism. That is not desirable.

If you are overweight and not losing weight adequately with this program, it may be because of medications such as insulin or sulfonylureas, which may need to be tapered or discontinued under physician supervision. Talk with your doctor.

Vegan or Flexitarian

The menus here are largely vegan with the option of including a small amount of animal products a few times a week. If you desire to include animal products in your diet, I strongly recommend that they only be used as a condiment to flavor a dish, not as a main part of the meal. In other words, do not have a whole piece of chicken in one meal, but instead use a piece of chicken to flavor a soup that may then be consumed over three to five days. If you're adding some shrimp or salmon to a salad or vegetable dish, use only a very small amount, such as one to two ounces. This nutritarian diet can be either vegan or flexitarian (minimal animal products), but don't sabotage the results with too many animal products.

Lastly, remember that eating this healthfully is the most delicious and pleasurable way to eat. As you gain experience with using whole, natural foods to revamp your health, you will realize that you can eat slower, chew well, and savor each bite. Nature is the best doctor and the best chef. Think about the good you are doing for yourself when you eat naturally nutrient-rich foods. When you purchase high-quality, fresh food, you get high-quality taste and a high-quality body.

The Menus

Week 1

DAY 1

Breakfast

    Soaked Oats and Blueberries*

Lunch

    Salad with romaine lettuce, mixed greens, and assorted vegetables topped with white beans

    Caesar Salad Dressing/Dip*

    Fresh fruit (apple)

Dinner

    Steamed asparagus with Sesame Ginger Sauce*

    Golden Austrian Cauliflower Cream Soup*

    Fresh fruit (melon)

DAY 2

Breakfast

    Fruit and nut bowl (assorted fresh and/or frozen fruit topped with nuts and seeds)

Lunch

    Romaine, spinach, and watercress salad topped with chickpeas

    Russian Fig Dressing/Dip*

    Golden Austrian Cauliflower Cream Soup*

    Fresh fruit (orange)

Dinner

    Raw vegetables

    Fresh Tomato Salsa*

    Simple Bean Burgers*

    Great Greens*

    Peach Sorbet*

Note: An asterisk indicates the recipe is included in this book.

DAY 3

Breakfast

Blended Mango Salad*

Walnuts

Lunch

Raw vegetables

Russian Fig Dressing/Dip*

Black Bean Lettuce Bundles*

Fresh fruit (two kiwis)

Dinner

Seasoned edamame

Roasted Vegetable Salad with Baked Tofu or Salmon*

Mixed berries

DAY 4

Breakfast

Quick Banana Oat Breakfast to Go*

Lunch

Mixed greens salad with assorted vegetables and flavored vinegar

Savory Portobello Mushrooms with Chickpeas*

Fresh fruit (grapes)

Dinner

Raw vegetables

Island Black Bean Dip*

Dr. Fuhrman's Famous Anticancer Soup*

Fresh fruit or Banana Walnut Ice Cream*

DAY 5

Breakfast

> Fruit and nut bowl (assorted fresh and/or frozen fruit
> topped with nuts or seeds)

Lunch

> Raw vegetables
> Island Black Bean Dip*
> Dr. Fuhrman's Famous Anticancer Soup*
> Fresh fruit (apple slices and cinnamon)

Dinner

> Romaine, spinach, and watercress salad
> Tofu Ranch Dressing/Dip*
> Eggplant Roll Ups*
> Fresh fruit (berries)

DAY 6

Breakfast

> Blue Apple Nut Oatmeal*

Lunch

> Salad with mixed greens and veggies
> Tofu Ranch Dressing/Dip*
> Yum Good Beans*
> Fresh fruit (strawberries)

Dinner

> Raw veggies
> Herbed White Bean Hummus*
> French Lentil Soup*
> Mango Coconut Sorbet*

DAY 7

Breakfast

Soaked Oats and Blueberries*

Lunch

Mixed greens salad with assorted vegetables and flavored vinegar

French Lentil Soup*

Fresh fruit (Grapes)

Dinner

The Big Veggie Stir-Fry*

Herbed Barley and Lentils*

Fresh fruit (apple slices with cinnamon)

Week 2

DAY 1

Breakfast

Fruit and nut bowl

Lunch

Romaine and arugula salad with assorted vegetables

Creamy Blueberry Dressing*

Vegetable Burrito*

Fresh fruit (pear)

Dinner

Raw vegetables

Choice of healthy dips

Easy Bean and Vegetable Chili*

Strawberries dusted with cocoa powder

DAY 2

Breakfast

    Blue Apple Nut Oatmeal*

Lunch

    Romaine and mixed greens salad and flavored vinegar

    Easy Bean and Vegetable Chili*

    Fresh fruit (blueberries)

Dinner

    Mushroom Soup Provencal*

    Choice of cooked vegetable with Almond Tomato Sauce*

    Fresh fruit (cherries)

DAY 3

Breakfast

    Green Gorilla Blended Salad*

Lunch

    Raw vegetables

    Choice of healthy dips

    Mushroom Soup Provencal*

    Fresh fruit (watermelon slices)

Dinner

    Steamed artichoke

    Green Velvet Dressing/Dip*

    Mediterranean Bean and Kale Sauté*

DAY 4

Breakfast

Blue Apple Nut Oatmeal*

Lunch

Romaine, spinach, and watercress salad

Thousand Island Dressing*

Speedy Vegetable Wrap*

Dinner

Raw vegetables

Fresh Tomato Salsa*

Pistachio-Crusted Tempeh with Balsamic Marinade
and Shiitake Mushrooms*

Braised Baby Bok Choy*

Blueberry Cobbler*

DAY 5

Breakfast

Tropical Fruit Salad*

Nut or seed topping

Lunch

Southern-Style Mixed Greens*

Fresh fruit (apple)

Dinner

Fast Mexican Black Bean Soup*

Swiss Chard with Garlic and Lemon*

Fresh fruit (melon)

DAY 6

Breakfast

    Fruit and nut bowl

Lunch

    Fast Mexican Black Bean Soup*

    Asparagus Polonaise*

    Fresh fruit (orange)

Dinner

    Romaine salad with assorted vegetables

    Caesar Salad Dressing/Dip*

    No-Meat Balls*

    Low-sodium marinara sauce

    Garlicky Zucchini*

    Fresh fruit (pear)

DAY 7

Breakfast

    Quick Banana Oat Breakfast to Go*

Lunch

    Mixed greens salad

    Flavored vinegar or choice of healthy dressings

    Steamed vegetable of choice with Red Lentil Sauce*

    Fresh fruit (papaya)

Dinner

    Raw vegetables

    Choice of healthy dips

    Thai Vegetable Curry*

    Strawberry Pineapple Sorbet*

The Recipes

BREAKFAST

Blue Apple Nut Oatmeal
Serves: 2

  1⅔ cups water
  ¼ teaspoon cinnamon
  ⅓ cup old-fashioned rolled oats
  1 cup frozen blueberries
  1 apple, chopped or grated
  2 tablespoons chopped walnuts

In a saucepan, combine water with cinnamon and oats. Simmer about 5 minutes, until oatmeal is creamy. Stir in blueberries, apples, and nuts.

PER SERVING: CALORIES 174; PROTEIN 5g; CARBOHYDRATES 29g; TOTAL FAT 6.3g; SATURATED FAT 0.6g; SODIUM 9mg, FIBER 7.6g; BETA-CAROTENE 41ug; VITAMIN C 5mg; CALCIUM 27mg; IRON 1.1mg; FOLATE 20ug; MAGNESIUM 57mg; ZINC 0.7mg; SELENIUM 5.0ug

Quick Banana Oat Breakfast to Go
Serves: 2

  2 cups frozen blueberries
  ½ cup old-fashioned rolled oats
  ⅓ cup pomegranate juice
  2 tablespoons chopped walnuts
  1 tablespoon raw sunflower seeds
  1 banana, sliced

In a cereal bowl, combine all ingredients. Heat in the microwave for 3 minutes.

Note: On the go, combine all ingredients in a sealed container to eat later, either hot or cold.

PER SERVING: CALORIES 308; PROTEIN 6g; CARBOHYDRATES 54g; TOTAL FAT 9.6g; SATURATED FAT 1.0g; SODIUM 4mg; FIBER 8.7g; BETA-CAROTENE 61ug; VITAMIN C 9mg; CALCIUM 28mg; IRON 1.8mg; FOLATE 48ug; MAGNESIUM 105mg; ZINC 1.2mg; SELENIUM 10.6ug

Soaked Oats and Blueberries

Serves: 2

¼ cup unsweetened soy, hemp, or almond milk
1 cup old-fashioned rolled oats
½ cup water
1 cup blueberries (fresh or frozen)
1 green apple, chopped
1 tablespoon ground flaxseeds

Pour milk over raw oats, add enough water to cover the oats, and soak for at least 30 minutes, preferably overnight. Add berries, apples, and flaxseeds. Mix and serve.

PER SERVING: CALORIES 261; PROTEIN 18g; CARBOHYDRATES 51g; TOTAL FAT 5.4g; SATURATED FAT 0.6g; SODIUM 19mg; FIBER 8.7g; BETA-CAROTENE 152ug; VITAMIN C 10mg; CALCIUM 29mg; IRON 2.6mg; FOLATE 30ug; MAGNESIUM 137mg; ZINC 1.7mg; SELENIUM 16ug

Tropical Fruit Salad

Serves: 4

2 cups cubed pineapple
1 cup cubed mango
1 cup cubed papaya
2 oranges, peeled and sliced
1 banana, sliced
2 tablespoons unsweetened shredded coconut
2 cups curly leaf lettuce leaves

Toss fruit together. Add coconut and serve on top of lettuce.

PER SERVING: CALORIES 146; PROTEIN 2g; CARBOHYDRATES 36g; TOTAL FAT 1.3g; SATURATED FAT 0.8g; SODIUM 5mg; FIBER 5.3g; BETA-CAROTENE 848ug; VITAMIN C 102mg; CALCIUM 55mg; IRON 0.6mg; FOLATE 76ug; MAGNESIUM 34mg; ZINC 0.3mg; SELENIUM 1.5ug

GREEN SMOOTHIES

Blended Mango Salad

Serves: 2

> 1 large, ripe (preferably chilled) mango
> 1 cup chopped spinach
> 4 cups chopped romaine lettuce
> ¼ cup unsweetened soy, hemp, or almond milk

Peel and chop mango and place in food processor or high-powered blender. Add spinach and half of lettuce. Blend. Add milk and the remaining lettuce. Blend until creamy.

PER SERVING: CALORIES 105; PROTEIN 4g; CARBOHYDRATES 23g; TOTAL FAT 1.2g; SATURATED FAT 0.2g; SODIUM 39mg; FIBER 4.8g; BETA-CAROTENE 5316ug; VITAMIN C 60mg; CALCIUM 74mg; IRON 1.9mg; FOLATE 201ug; MAGNESIUM 44mg; ZINC 0.5mg; SELENIUM 2.6ug

Green Gorilla Blended Salad

Serves: 2

> ½ avocado
> 1 banana
> 5 ounces baby romaine lettuce
> 5 ounces organic baby spinach

In a food processor or high-powered blender, blend avocado with the banana. Then add lettuce and spinach. Process until smooth.

PER SERVING: CALORIES 172; PROTEIN 5g; CARBOHYDRATES 24g; TOTAL FAT 8.3g; SATURATED FAT 1.6g; SODIUM 64mg; FIBER 8.8g; BETA-CAROTENE 6512ug; VITAMIN C 55mg; CALCIUM 104mg; IRON 2.9mg; FOLATE 272ug; MAGNESIUM 100mg; ZINC 0.9mg; SELENIUM 1.6ug

DIPS, DRESSINGS, AND SAUCES

Dips and Dressings

Caesar Salad Dressing/Dip

Serves: 3

> 3 cloves garlic, roasted
> 3 tablespoons raw, un-hulled sesame seeds
> ½ cup unsweetened soy, hemp, or almond milk
> ⅓ cup raw cashews
> 1 tablespoon fresh lemon juice
> 1½ tablespoons nutritional yeast
> 1½ teaspoons Dijon mustard
> Black pepper

Break the garlic cloves apart and leave the papery skins on. Roast them on a baking pan at 350°F until soft (about 20 minutes). In a flat pan, toast the sesame seeds, shaking for 3 minutes, and put aside. Skin the roasted garlic and place it in a food processor or high-powered blender along with the milk, cashews, lemon juice, nutritional yeast, and mustard. Sprinkle on a dash of black pepper. Blend until creamy and smooth. Drizzle the dressing over a tossed salad and top with the toasted sesame seeds. If using as a dip, mix the sesame seeds into the dip.

PER SERVING: CALORIES 266; PROTEIN 11g; CARBOHYDRATES 16g; TOTAL FAT 19.6g; SATURATED FAT 3.5g; SODIUM 111mg; FIBER 3.5g; BETA-CAROTENE 147ug; VITAMIN C 4mg; CALCIUM 125mg; IRON 4.2mg; FOLATE 176ug; MAGNESIUM 122mg; ZINC 2.7mg; SELENIUM 7.6ug

Creamy Blueberry Dressing

Serves: 4

2 cups fresh, or frozen and thawed, blueberries
½ cup pomegranate juice
⅓ cup raw cashews
6 walnut halves
3 tablespoons Wild Blueberry Vinegar or white wine vinegar

Blend all ingredients in a food processor or high-powered blender until smooth and creamy.

PER SERVING: CALORIES 165; PROTEIN 4g; CARBOHYDRATES 19g; TOTAL FAT 9.0g; SATURATED FAT 1.4g; SODIUM 5mg; FIBER 2.7g; BETA-CAROTENE 22ug; VITAMIN C 5mg; CALCIUM 15mg; IRON 1.5mg; FOLATE 11ug; MAGNESIUM 57mg; ZINC 1.1mg; SELENIUM 3.6ug

Fresh Tomato Salsa

Serves: 4

2 fresh tomatoes, chopped
1 small red onion, finely chopped
2 scallions, finely chopped
2 cloves garlic, finely chopped
½ jalapeno chili pepper, seeded and minced
3 tablespoons chopped cilantro
3 tablespoons fresh lime juice

Stir together all ingredients. Serve immediately, or refrigerate in an airtight, covered container for up to 3 days.

PER SERVING: CALORIES 15; PROTEIN 1g; CARBOHYDRATES 4g; TOTAL FAT 0.1g; SODIUM 4mg; FIBER 0.8g; BETA-CAROTENE 228ug; VITAMIN C 8mg; CALCIUM 12mg; IRON 0.2mg; FOLATE 12ug; MAGNESIUM 7mg; ZINC 0.1mg; SELENIUM 0.2ug

Garbanzo Guacamole

Serves: 3

1½ cups cooked garbanzo beans or 1 (15-ounce) can, no salt
added, drained
2 cloves garlic
1 tablespoon lemon juice
1 avocado, peeled and cubed
1½ fresh green chili peppers, minced
1 cup chopped tomato
¾ cup chopped green onions
1 teaspoon Bragg Liquid Aminos or low-sodium soy sauce
2 tablespoons chopped cilantro

In a food processor, puree beans and garlic with lemon juice. Add
avocado and chili peppers, pulsing until mixture is fairly smooth with
some small chunks. Remove to bowl and stir in tomato, green onions,
liquid aminos, and cilantro. Serve with raw vegetables, or refrigerate
in an airtight container for up to 4 days.

PER SERVING: CALORIES 261; PROTEIN 11g; CARBOHYDRATES 35g; TOTAL FAT 11g; SATURATED FAT 1g;
SODIUM 87mg; FIBER 13.1g; BETA-CAROTENE 918ug; VITAMIN C 113mg; CALCIUM 87mg; IRON 3.6mg; FOLATE 228ug;
MAGNESIUM 82mg; ZINC 2mg; SELENIUM 3.3

Green Velvet Dressing/Dip .

Serves: 4

¾ cup water
½ cup fresh lemon juice
½ cup un-hulled, raw sesame seeds or raw tahini
⅓ cup raw cashews
¼ cup chopped fresh parsley
¼ cup chopped fresh dill
2 tablespoons VegiZest or other no-salt seasoning blend
½ tablespoon chopped fresh tarragon or ½ teaspoon dried
1 teaspoon Bragg Liquid Aminos or low-sodium soy sauce
2 cloves garlic

Blend all ingredients in a food processor or high-powered blender until smooth. Serve, or refrigerate in an airtight container for up to 4 days.

PER SERVING: CALORIES 191; PROTEIN 6g; CARBOHYDRATES 13g; TOTAL FAT 14.3g; SATURATED FAT 2.3g; SODIUM 88mg; FIBER 2.7g; BETA-CAROTENE 791ug; VITAMIN C 22mg; CALCIUM 194mg; IRON 4mg; FOLATE 36ug; MAGNESIUM 98mg; ZINC 2mg; SELENIUM 2.6ug

Herbed White Bean Hummus

Serves: 2

1⅔ cups cooked white beans or 1 (15-ounce) can, no salt added,
  drained
1 tablespoon lemon juice
2 tablespoons un-hulled, raw sesame seeds
2 tablespoons red wine vinegar
½ teaspoon Dijon mustard
2 tablespoons water
¼ cup chopped fresh basil
2 tablespoons fresh thyme

Place beans, lemon juice, seeds, vinegar, mustard, and water in a high-powered blender or food processor. Blend until smooth. Add basil and thyme and pulse very briefly. Do not overprocess herbs; they should be visible in small pieces. Serve, or refrigerate in an airtight container for up to 4 days.

PER SERVING: CALORIES 180; PROTEIN 10g; CARBOHYDRATES 30g; TOTAL FAT 2.7g; SATURATED FAT 0.4g; SODIUM 23mg; FIBER 7.1g; BETA-CAROTENE 118ug; VITAMIN C 4mg; CALCIUM 149mg; IRON 4.9mg; FOLATE 92ug; MAGNESIUM 87mg; ZINC 1.9mg; SELENIUM 2.4ug

Island Black Bean Dip

Serves: 4

1⅔ cups cooked black beans or 1 (15-ounce) can, no salt added, drained

2 teaspoons no-salt salsa

¼ cup minced scallions

1½ tablespoons Riesling Reserve Vinegar or other fruity flavored vinegar

2 tablespoons MatoZest or other no-salt seasoning blend

2 tablespoons minced red onion

½ cup finely diced mango

¼ cup diced red bell pepper

1 tablespoon fresh minced cilantro or parsley, for garnish

Remove ¼ cup of the black beans and set aside. Place remaining beans in a blender or food processor. Add salsa, scallions, vinegar, and seasoning mix. Puree until relatively smooth. Adjust seasonings to taste. Transfer to a bowl and add the reserved black beans, red onion, mango, and red bell pepper. Mix well and chill for 1 hour. Garnish with cilantro or parsley. Serve with unsalted, oil-free baked pita or raw veggies, or refrigerate in an airtight container for up to 4 days.

PER SERVING: CALORIES 117; PROTEIN 7g; CARBOHYDRATES 23g; TOTAL FAT 0.5g; SATURATED FAT 0.1g; SODIUM 40mg; FIBER 6.5g; BETA-CAROTENE 926ug; VITAMIN C 22mg; CALCIUM 28mg; IRON 1.9mg; FOLATE 109ug; MAGNESIUM 51mg; ZINC 0.8mg; SELENIUM 1ug

Russian Fig Dressing/Dip

Serves: 2

5 tablespoons no- or low-salt pasta sauce

3 tablespoons raw almond butter or 1.5 ounces raw almonds

2 tablespoons raw sunflower seeds

3 tablespoons Black Fig Vinegar or balsamic vinegar

Blend all ingredients in a food processor or high-powered blender until smooth. Serve, or refrigerate in an airtight container for up to 4 days.

PER SERVING: CALORIES 206; PROTEIN 5g; CARBOHYDRATES 15g; TOTAL FAT 14.2g; SATURATED FAT 1.3g; SODIUM 18mg; FIBER 4.0g; BETA-CAROTENE 151ug; VITAMIN C 5mg; CALCIUM 67mg; IRON 1.7mg; FOLATE 30ug; MAGNESIUM 87mg; ZINC 1.1mg; SELENIUM 6.3ug

Thousand Island Dressing

Serves: 4

½ cup raw almonds

½ cup raw cashews

½ cup unsweetened soy, hemp, or almond milk

2 tablespoons balsamic vinegar

2 tablespoons lemon juice

1 tablespoon fresh dill or 1 teaspoon dried

1 teaspoon onion powder or granules

1 bulb or 8 cloves roasted garlic

3 tablespoons tomato paste

½ cucumber, peeled and cut into two equal portions

¼ cup finely chopped onion

In a food processor or high-powered blender, blend together the nuts, milk, vinegar, lemon juice, dill, onion powder, garlic, tomato paste, and half of the cucumber until smooth.

Remove from blender. Finely chop remainder of cucumber, combine with onions, and fold into the dip. Serve, or refrigerate in an airtight container for up to 4 days.

PER SERVING: CALORIES 246; PROTEIN 9g; CARBOHYDRATES 18g; TOTAL FAT 18g; SATURATED FAT 2.3g; SODIUM 123mg; FIBER 3.7g; BETA-CAROTENE 239ug; VITAMIN C 8mg; CALCIUM 74mg; IRON 3mg; FOLATE 30ug; MAGNESIUM 114mg; ZINC 1.8mg; SELENIUM 4.9ug

Tofu Ranch Dressing/Dip

Serves: 4

6 ounces silken tofu

3 dates, pitted

1 clove garlic

¼ cup finely chopped green onion

3 tablespoons water

2 tablespoons lemon juice

1½ tablespoons Italian seasoning

1 tablespoon chopped fresh parsley or 1 teaspoon dried

1 tablespoon chopped fresh dill or 1 teaspoon dried

1 teaspoon Bragg Liquid Aminos or low-sodium soy sauce

Cayenne pepper

In a high-powered blender or food processor, combine all ingredients and process until smooth. Serve, or refrigerate in an airtight container for up to 5 days.

Note: Use as a dressing, dip, spread, or mayonnaise substitute in your favorite recipes.

PER SERVING: CALORIES 61; PROTEIN 4g; CARBOHYDRATES 8g; TOTAL FAT 2.3g; SATURATED FAT 0.3g; SODIUM 56mg; FIBER 1.6g; BETA-CAROTENE 167ug; VITAMIN C 7mg; CALCIUM 187mg; IRON 3.3mg; FOLATE 19ug; MAGNESIUM 23mg; ZINC 0.5mg; SELENIUM 4.2ug

Sauces for Cooked Vegetables

Almond Tomato Sauce

Serves: 4

> ½ cup raw almonds
> 1 large clove garlic
> 2 medium tomatoes
> ¼ teaspoon sweet paprika
> 1 tablespoon red wine vinegar
> ¼ cup unsweetened soy, hemp, or almond milk
> 2 tablespoons fresh basil

Process almonds in a food processor until finely ground. Add garlic, tomatoes, paprika, vinegar, and milk and continue to process until smooth. Add basil and pulse until mixed in. Serve over steamed or water-sautéed vegetables. Or use as a topping for a mixture of roasted red peppers and steamed snow peas.

PER SERVING: CALORIES 135; PROTEIN 7g; CARBOHYDRATES 9g; TOTAL FAT 9.6g; SATURATED FAT 0.8g; SODIUM 18mg; FIBER 3.0g; BETA-CAROTENE 410ug; VITAMIN C 10mg; CALCIUM 64mg; IRON 1.2mg; FOLATE 18ug; MAGNESIUM 63mg; ZINC 0.8mg; SELENIUM 2.0ug

Red Lentil Sauce

Serves: 4

½ cup red lentils
1 medium onion, chopped
1 clove garlic, chopped
1 cup carrot juice
1 tablespoon VegiZest or other no-salt seasoning blend
1 teaspoon cumin
½ teaspoon balsamic vinegar
½ cup water
¼ cup chopped pecans

Add lentils, onions, garlic, and carrot juice to a saucepan over high heat. Bring to a boil, cover, and simmer until the lentils are soft and pale (20–30 minutes). Add more carrot juice if needed.

Put the cooked lentil mixture into a food processor or blender with VegiZest, cumin, and balsamic vinegar and blend to a smooth puree. Add additional water if it is too thick. Serve with steamed broccoli, cauliflower, or other vegetables. Top with chopped pecans.

PER SERVING: CALORIES 176; PROTEIN 8g; CARBOHYDRATES 25g; TOTAL FAT 5.4g; SATURATED FAT 0.5g; SODIUM 31mg; FIBER 9.1g; BETA-CAROTENE 5797ug; VITAMIN C 10mg; CALCIUM 45mg; IRON 2.5mg; FOLATE 124ug; MAGNESIUM 49mg; ZINC 1.6mg; SELENIUM 3ug

Sesame Ginger Sauce

Serves: 4

½ cup tahini
2 tablespoons fresh lemon juice
1 teaspoon white miso
1 tablespoon finely grated ginger
2 pitted dates
1 clove garlic, crushed
Freshly ground pepper
⅔ cup water

Blend all ingredients in a high-powered blender or food processor. Additional water can be added to achieve desired consistency. Serve with steamed or water-sautéed vegetables.

This sauce goes well with bok choy, asparagus, or kale.

PER SERVING: CALORIES 190; PROTEIN 6g; CARBOHYDRATES 13g; TOTAL FAT 14.5g; SATURATED FAT 2g; SODIUM 63mg; FIBER 3.2g; VITAMIN C 4mg; CALCIUM 131mg; IRON 0.8mg; FOLATE 31ug; MAGNESIUM 32mg; ZINC 1.4mg; SELENIUM 0.2ug

SOUPS

Dr. Fuhrman's Famous Anticancer Soup

Serves: 10

6–8 medium zucchini
1½ cups mixed dry split peas, lentils, and beans
6 cups water
5 pounds carrots or 5–6 cups carrot juice
2 bunches organic celery or 2 cups celery juice
4 medium onions
3 leek stalks, sliced lengthwise to unravel and rinse
2 bunches kale, collards, or other greens, chopped, tough stems
　　and center ribs cut off and discarded
½ cup raw cashews
8 ounces mushrooms (shiitake, cremini, and/or oyster), chopped
2 tablespoons VegiZest or other no-salt seasoning blend
1 teaspoon Mrs. Dash or other no-salt spice blend

Place the zucchini, beans, and water in a very large pot over low heat and cover. Juice carrots and celery and add to pot. Blend onions, leeks, and kale in a high-powered blender with some of the soup liquid. Remove the partially cooked zucchini from the pot. Add it and the cashews to the blended onions, leeks, and kale. Blend mix in batches until thick and creamy and return to the pot. Add the seasonings and mix thoroughly. Add the chopped mushrooms and simmer until beans are soft (about one hour).

Note: Juice carrots and celery in a juice extractor. Freshly juiced organic carrots maximize the flavor of this soup.

PER SERVING: CALORIES 322; PROTEIN 16g; CARBOHYDRATES 61g; TOTAL FAT 4.6g; SATURATED FAT 0.8g; SODIUM 132mg; FIBER 12g; BETA-CAROTENE 26638ug; VITAMIN C 160mg; CALCIUM 241mg; IRON 5.7mg; FOLATE 254ug; MAGNESIUM 154mg; ZINC 3.0mg; SELENIUM 7.0ug

Fast Mexican Black Bean Soup

Serves: 5

> 2 (15-ounce) cans black beans, no salt added
> 2 cups frozen mixed vegetables
> 2 cups frozen corn
> 2 cups frozen chopped broccoli florets
> 2 cups carrot juice, fresh or bottled
> 1 cup water
> 1 cup prepared black bean soup, no or low salt
> ¼ cup chopped cilantro
> 1 teaspoon chili powder, or to taste
> 1 cup chopped fresh tomatoes
> ½ cup chopped scallions
> ¼ cup raw pumpkin seeds

Combine black beans, mixed vegetables, corn, broccoli, tomatoes, carrot juice, water, soup, cilantro, and chili powder in a soup pot. Bring to a boil and simmer on low for 20 minutes.

Serve topped with extra tomatoes, scallions, and pumpkin seeds.

PER SERVING: CALORIES 412; PROTEIN 21g; CARBOHYDRATES 68g; TOTAL FAT 10.3g; SATURATED FAT 1.7g; SODIUM 84mg; FIBER 20.4g; BETA-CAROTENE 9564ug; VITAMIN C 64mg; CALCIUM 129mg; IRON 5.8mg; FOLATE 314ug; MAGNESIUM 189mg; ZINC 3.2mg; SELENIUM 5.1ug

French Lentil Soup

Serves: 8

8 cups water

2 cups carrot juice, fresh or bottled

2 (15-ounce) cans no-salt added diced tomatoes,
   or 3 cups diced fresh tomatoes

4 tablespoons VegiZest or other no-salt seasoning

2 onions, chopped

3 ribs celery, chopped

2 carrots, sliced

6 cloves garlic, minced

1 tablespoon dried oregano

1½ teaspoons dried basil or 2 tablespoons chopped fresh

1 teaspoon ground coriander

1 teaspoon ground cumin

1 teaspoon dried thyme or 1 tablespoon chopped fresh

1 pound dry French (green) lentils, rinsed

1 red bell pepper, chopped

½ green bell pepper, chopped

2 small zucchini, chopped

1 (10-ounce) box frozen lima beans

1 bunch kale, tough stems removed, finely chopped

3 tablespoons Riesling Reserve Vinegar

½ cup raw cashews

½ cup raw filberts

Put all the ingredients except for the nuts in a large soup pot. Bring to a simmer and cook until lentils are tender (about 1 hour). Put 2 cups or more of the soup into a food processor or high-powered blender, add nuts, blend until smooth, and stir back into soup. Serve.

PER SERVING: CALORIES 422; PROTEIN 25g; CARBOHYDRATES 63g; TOTAL FAT 10.4g; SATURATED FAT 1.4g; SODIUM 99mg; FIBER 23.3g; BETA-CAROTENE 9608ug; VITAMIN C 73mg; CALCIUM 135mg; IRON 7.1mg; FOLATE 333ug; MAGNESIUM 152mg; ZINC 4.1mg; SELENIUM 7.5ug

Golden Austrian Cauliflower Cream Soup

Serves: 4

  1 head cauliflower, cut into florets
  3 carrots, coarsely chopped
  1 cup coarsely chopped organic celery
  2 leeks, coarsely chopped
  2 tablespoons VegiZest or other no-salt seasoning blend
  1 teaspoon Mrs. Dash or other no-salt spice blend
  2 cups carrot juice, fresh or bottled
  4 cups water
  2 cloves garlic, minced
  ½ teaspoon nutmeg
  1 cup raw cashews
  5 cups chopped organic baby spinach

Place cauliflower, carrots, celery, leeks, VegiZest, Mrs. Dash, carrot juice, water, garlic, and nutmeg in a large soup pot, cover, and simmer until cauliflower is soft (about 15 minutes). Transfer ⅔ of the soup (vegetables and stock) to a blender or food processor, add cashews, and blend until smooth and creamy. Return blended mixture to the soup pot, stir in spinach, and simmer until spinach is wilted. Serve.

PER SERVING: CALORIES 354; PROTEIN 13g; CARBOHYDRATES 46g; TOTAL FAT 16.7g; SATURATED FAT 3.4g; SODIUM 202mg; FIBER 9.1g; BETA-CAROTENE 18003ug; VITAMIN C 102mg; CALCIUM 176mg; IRON 5.8mg; FOLATE 233ug; MAGNESIUM 182mg; ZINC 3mg; SELENIUM 6.8ug

Mushroom Soup Provencal

Serves: 8

> 1⅓ cups dried white, Northern, navy, or cannellini beans, rinsed, or 2 (15-ounce) cans, no salt added, drained
> 4 cups water (if using canned beans, decrease water to 1 cup)
> 3 cups unsweetened soy, hemp, or almond milk
> 5 cups carrot juice, fresh or bottled
> 2 carrots, coarsely chopped
> 1 (10-ounce) package frozen corn
> 1 cup chopped organic celery
> 3 leeks, sliced lengthwise to unravel and rinse, then cut in ½-inch slices
> 2 pounds mixed fresh mushrooms (button, shiitake, or cremini), sliced ¼-inch thick
> 2 cloves garlic, minced or pressed
> 2 medium onions, chopped
> 2 teaspoons herbes de Provence
> 4 tablespoons VegiZest or other no-salt seasoning blend
> ¼ cup raw cashews
> 1 tablespoon lemon juice
> 1 tablespoon chopped fresh thyme or 1 teaspoon dried
> 2 teaspoons chopped fresh rosemary or ½ teaspoon dried
> 1 (6-ounce) bag baby spinach, chopped
> ¼ cup chopped fresh parsley

In a large bowl, submerge beans 1 inch below water and soak them overnight. Or in a soup pot, bring them to a rapid boil, cover, and set aside for 1 hour.*

Drain the beans. In a large soup pot, add 4 cups water and beans and heat on low. Add 2½ cups of the milk and all of the carrot juice, carrots, corn, celery, leeks, mushrooms, garlic, onions, herbes de Provence, and VegiZest to the pot. Simmer until beans are tender (about 1 hour).

In a blender or food processor, blend cashews and reserved ½ cup milk. Add half of the soup and the lemon juice, thyme, and rosemary. Blend until smooth and creamy.

(continued on next page)

Return pureed soup back to the pot and add chopped spinach. Cook on low flame until heated through and spinach is wilted.

Serve, garnished with parsley.

* If the beans are not presoaked, increase cooking time and use 8–10 cups water.

PER SERVING: CALORIES 305; PROTEIN 18g; CARBOHYDRATES 53g; TOTAL FAT 5.1g; SATURATED FAT 0.9g; SODIUM 164mg; FIBER 11g; BETA-CAROTENE 17408ug; VITAMIN C 36mg; CALCIUM 186mg; IRON 6mg; FOLATE 188ug; MAGNESIUM 133mg; ZINC 2.4mg; SELENIUM 19.9ug

Quick and Creamy Vegetable Bean Soup
Serves: 8

> 4 cups low-sodium or no-salt-added vegetable broth
> 2 cups frozen broccoli florets
> 2 cups frozen chopped organic spinach
> 2 cups carrot juice
> 1 cup frozen chopped onions
> 4 (15-ounce) cans cannellini beans or other white beans, no salt, drained
> 3 fresh tomatoes, chopped
> 1 bunch fresh basil, chopped
> 4 tablespoons VegiZest or other no-salt seasoning blend
> 1 teaspoon garlic powder
> 1 teaspoon Italian seasoning
> ½ cup raw cashews
> ¼ cup pine nuts

In a soup pot, combine all ingredients except cashews and pine nuts. Cover and simmer for 30–40 minutes. Transfer ¼ of the soup mixture to a high-powered blender, add cashews and pine nuts, and blend. Return blended mixture back to the soup pot, and serve.

PER SERVING: CALORIES 324; PROTEIN 18g; CARBOHYDRATES 51g; TOTAL FAT 8.4g; SATURATED FAT 1.2g; SODIUM 87mg; FIBER 10g; BETA-CAROTENE 7310ug; VITAMIN C 73mg; CALCIUM 165mg; IRON 7mg; FOLATE 151ug; MAGNESIUM 136mg; ZINC 2.8mg; SELENIUM 5.1ug

Tomato Bisque
Serves: 8

3 cups carrot juice

1½ pounds fresh tomatoes, chopped, or 1 (28-ounce) can San Marzano tomatoes

¼ cup chopped sun-dried tomatoes

2 organic celery ribs, chopped

1 small onion, chopped

1 leek, chopped

1 large shallot, chopped

3 cloves garlic, chopped

2 tablespoons VegiZest or other no-salt seasoning blend

1 teaspoon crumbled dried thyme

1 small bay leaf

½ cup raw cashews

¼ cup chopped fresh basil

1 (5-ounce) bag organic baby spinach

In a large saucepan, add carrot juice, fresh and sun-dried tomatoes, celery, onion, leek, shallot, garlic, VegiZest, thyme, and bay leaf. Simmer for 30 minutes. Discard bay leaf. With a slotted spoon, remove 2 cups of vegetables and set aside. Transfer remaining soup to a food processor, add cashews, and puree until smooth. Return puree to pot and add soup vegetables to make a creamy bisque with chunks of vegetables. Stir in basil and spinach and let them wilt. Serve.

PER SERVING: CALORIES 292; PROTEIN 15g; CARBOHYDRATES 45g; TOTAL FAT 11g; SATURATED FAT 1.9g; SODIUM 206mg; FIBER 9g; BETA-CAROTENE 20201ug; VITAMIN C 58mg; CALCIUM 180mg; IRON 5.3mg; FOLATE 154ug; MAGNESIUM 143mg; ZINC 2.2mg; SELENIUM 4.5ug

MAIN DISHES

Asparagus Polonaise

Serves: 4

> 2 bunches asparagus
> 7 ounces (or half 15.5 ounce package) soft tofu
> 1 tablespoon lemon juice
> 2 pitted dates
> 1 clove garlic, chopped
> ½ cup fresh chopped parsley
> ½ cup unsweetened soy, hemp, or almond milk

Cut tough ends off asparagus and steam until tender (about 8 minutes). In a food processor or high-powered blender, blend tofu, lemon juice, dates, garlic, parsley, and milk until smooth. Pour over asparagus, and serve.

PER SERVING: CALORIES 126; PROTEIN 8g; CARBOHYDRATES 21g; TOTAL FAT 1.8g; SATURATED FAT 0.3g; SODIUM 30mg; FIBER 5.8g; BETA-CAROTENE 1508ug; VITAMIN C 25mg; CALCIUM 119mg; IRON 6.3mg; FOLATE 148ug; MAGNESIUM 58mg; ZINC 1.7mg; SELENIUM 12.1ug

The Big Veggie Stir-Fry

Serves: 4

> 1 tablespoon minced fresh ginger
> 2 cloves garlic, minced
> 2 tablespoons tahini
> ¼ teaspoon hot pepper flakes
> ¼ cup pineapple juice
> 2 tablespoons rice vinegar
> 2 teaspoons reduced-sodium soy sauce

1½ cups water
1 (8-ounce) tofu, cut into 1-inch cubes
5 cups broccoli florets
2 cups chopped bok choy
1 zucchini, chopped
1 red bell pepper, chopped
4 cups shredded cabbage
2 stems lemongrass, thinly sliced
2 teaspoons cornstarch
4 scallions, chopped
2 tablespoons sesame seeds

In a small bowl, combine ginger, garlic, tahini, hot pepper flakes, pineapple juice, rice vinegar, soy sauce, and ¼ cup water. Set aside.

Rub a large wok with oil and stir-fry tofu over high heat for 5 minutes or until slightly browned. Transfer to a bowl and tent to keep warm.

To the wok, add broccoli, bok choy, zucchini, red pepper, cabbage, and lemongrass along with ¾–1 cup of water. Cover and cook over medium-high heat, stirring occasionally, until crisp but tender (about 10 minutes). Stir tofu into veggie mixture in wok and add sauce mixture. Stir-fry until heated through (about 2 minutes).

In a small bowl, combine cornstarch and 3 tablespoons water and stir into wok. Cook until thickened (2–3 minutes). Additional water may be added to adjust consistency. Top with scallions and sesame seeds. Serve over Herbed Barley and Lentils.

Note: Substitute 1 chicken breast, cut into strips, shrimp, or scallops for tofu if you wish.

PER SERVING: CALORIES 210; PROTEIN 11g; CARBOHYDRATES 29g; TOTAL FAT 8.0g; SATURATED FAT 1g; SODIUM 177mg; FIBER 7.7g; BETA-CAROTENE 1271ug; VITAMIN C 187mg; CALCIUM 223mg; IRON 3.8mg; FOLATE 171ug; MAGNESIUM 93mg; ZINC 2.0mg; SELENIUM 10.1ug

Black Bean Lettuce Bundles

Serves: 4

    1½ cups cooked black beans or 1 (15-ounce) can, no salt added
    ½ large avocado
    ½ green bell pepper, chopped
    3 scallions, chopped
    ⅓ cup chopped fresh cilantro
    2 tablespoons lime juice
    1 teaspoon cumin
    2 cloves garlic, minced
    ⅓ cup mild salsa, low sodium
    8 large romaine lettuce leaves

With a fork, mash beans and avocado together. In a medium bowl, combine the mash with remaining ingredients, except lettuce, and mix. Place approximately ¼ cup of filling in center of lettuce leaf, roll like a tortilla, and eat.

PER SERVING: CALORIES 146; PROTEIN 8g; CARBOHYDRATES 22g; TOTAL FAT 4.0g; SATURATED FAT 0.6g; SODIUM 15mg; FIBER 9.4g; BETA-CAROTENE 2352ug; VITAMIN C 36mg; CALCIUM 61mg; IRON 2.4mg; FOLATE 207ug; MAGNESIUM 67mg; ZINC 1.1mg; SELENIUM 1.3brasieg

Braised Baby Bok Choy

Serves: 4

8 baby bok choy or 3 regular bok choy
1 tablespoon Bragg Liquid Aminos or low-sodium soy sauce
1 tablespoon un-hulled sesame seeds, lightly toasted

Cover bottom of large skillet with ½ inch water. Cut baby bok choy in half lengthwise, or cut regular bok choy into chunks, and add to the water. Drizzle with liquid aminos. Cover and cook on high heat until bok choy is tender (about 6 minutes).

Lightly toast sesame seeds in a pan over medium heat for 3 minutes, shaking pan frequently.

Plate the bok choy and top with toasted sesame seeds.

PER SERVING: CALORIES 53; PROTEIN 5g; CARBOHYDRATES 4g; TOTAL FAT 1.7g; SATURATED FAT 0.2g; CHOLESTEROL 0.1mg; SODIUM 244mg; FIBER 3.5g; VITAMIN C 42mg; CALCIUM 157mg; IRON 1.8mg; FOLATE 83ug; MAGNESIUM 30mg; ZINC 0.8mg; SELENIUM 1.6ug

California Creamed Kale

Serves: 4

> 2 bunches kale, tough stems removed, chopped
> 1 cup raw cashews
> 1 cup unsweetened soy, almond, or hemp milk
> 4 tablespoons onion flakes
> 1 tablespoon VegiZest or no-salt seasoning blend (optional)

Place kale in a large steamer pot. Steam until soft (10–20 minutes).

Meanwhile, place remaining ingredients in a high-powered blender and blend until smooth.

Place steamed kale in colander and press with a clean dish towel to remove some of the excess water. Coarsely chop the kale, transfer it to a bowl, and mix it with the cream sauce.

Note: Sauce may be used with broccoli, spinach, or other steamed vegetables.

PER SERVING: CALORIES 279; PROTEIN 12g; CARBOHYDRATES 26g; TOTAL FAT 16.7g; SATURATED FAT 2.9g; SODIUM 79mg; FIBER 3.8g; BETA-CAROTENE 7060ug; VITAMIN C 90mg; CALCIUM 144mg; IRON 4.4mg; FOLATE 47ug; MAGNESIUM 144mg; ZINC 2.7mg; SELENIUM 10.6ug

Cauliflower, Spinach Mashed "Potatoes"

Serves: 4

6 cups cauliflower florets, fresh or frozen
6 cloves garlic, sliced
1 (10-ounce) bag fresh organic spinach
½ cup raw cashew butter
soy milk, as needed to thin
2 tablespoons VegiZest or other no-salt seasoning blend
¼ teaspoon nutmeg

In a vegetable steamer, steam cauliflower and garlic until tender (about 8 to 10 minutes). Drain, pressing out as much water as possible. Set aside.

Place spinach in steamer, steam until just wilted, and set aside.

With the "S" blade of a food processor, process cauliflower, garlic, and cashew butter until creamy and smooth. If it is too thick, add a small amount of soy milk, process, and check again. Add VegiZest and nutmeg, adjusting to taste. Mix pureed cauliflower with wilted spinach. Serve hot or warm.

PER SERVING: CALORIES 164; PROTEIN 9g; CARBOHYDRATES 18g; TOTAL FAT 8.5g; SATURATED FAT 1.7g; SODIUM 124mg; FIBER 5.7g; BETA-CAROTENE 4599ug; VITAMIN C 93mg; CALCIUM 116mg; IRON 3.8mg; FOLATE 234ug; MAGNESIUM 121mg; ZINC 1.7mg; SELENIUM 3.9ug

Easy Bean and Vegetable Chili
Serves: 6

> 1 pound firm tofu, frozen, defrosted
> 5 teaspoons chili powder, or to taste
> 1 teaspoon cumin
> 10 ounces frozen chopped onions
> 3 cups frozen broccoli, finely chopped
> 3 cups frozen cauliflower, finely chopped
> 3 cloves garlic
> 1 can pinto beans, no salt added, drained
> 1 can black beans, no salt added, drained
> 1 can red beans, no salt added, drained
> 1 (28-ounce) can diced tomatoes, no salt added,
>   or 2½ cups diced fresh tomatoes
> 1 (4.5-ounce) can chopped green chilies, mild
> 2½ cups corn, fresh or frozen
> 2 large zucchini, finely chopped

Squeeze excess water out of tofu and crumble. Place in a large pan
on high heat, add chili powder and cumin, and quickly brown. Add
onions, broccoli, cauliflower, garlic, beans, tomatoes, chilies, corn,
and zucchini, cover, and simmer for 2 hours. Serve hot.

PER SERVING: CALORIES 392; PROTEIN 29g; CARBOHYDRATES 66g; TOTAL FAT 5.4g; SATURATED FAT
0.4g; SODIUM 158 mg; FIBER 20.5 g; BETA-CAROTENE 1572 ug; VITAMIN C 121 mg; CALCIUM 263 mg; IRON 6.5
mg; FOLATE 371 ug; MAGNESIUM 188 mg; ZINC 2.6 mg; SELENIUM 7.5 ug

Eggplant Roll Ups

Serves: 4

2 tablespoons balsamic vinegar
2 tablespoons plus ¼ cup water
1 large eggplant, peeled and sliced into thin, flat, long strips
Nonstick cooking spray
1 red or green bell pepper, or ½ of each, diced
1 onion, chopped
2 cloves garlic, chopped
1 cup tomato sauce, no salt added
1 cup diced tomatoes
1 teaspoon Italian herb seasoning
1 teaspoon date sugar

Combine balsamic vinegar and 2 tablespoons water, add eggplant slices, and marinate for 1 hour. Drain eggplant and place in a baking pan that has been lightly coated with cooking spray. Bake at 350°F degrees until flexible (about 20 minutes).

Meanwhile, sauté pepper, onion, and garlic in ¼ cup water over medium heat for 5 minutes or until water is cooked down. Add tomato sauce, diced tomatoes, seasoning, and date sugar and simmer for 10 minutes. Take a strip of partially cooked eggplant, spread a layer of the tomato sauce mixture in the middle, and roll up. Repeat for all eggplant slices. Place in a baking dish, cover with the remaining tomato sauce mixture, and bake at 350°F for 30 minutes. Serve hot.

PER SERVING: CALORIES 96; PROTEIN 3g; CARBOHYDRATES 22g; TOTAL FAT 0.5g; SATURATED FAT 0.1g; SODIUM 22mg; FIBER 7.3g; BETA-CAROTENE 674ug; VITAMIN C 57mg; CALCIUM 54mg; IRON 1.7mg; FOLATE 59ug; MAGNESIUM 45mg; ZINC 0.6mg; SELENIUM 1.2ug

Garlicky Zucchini

Serves: 4

6 cups sliced zucchini
4 cloves garlic, minced
¼ teaspoon black pepper
2 tablespoons chopped lightly toasted pine nuts

Sauté zucchini and garlic in a small amount of water over medium-high heat until tender (about 3 minutes). Add more water if necessary to prevent scorching. Add black pepper and sprinkle with chopped pine nuts. Serve warm.

PER SERVING: CALORIES 45; PROTEIN 3g; CARBOHYDRATES 7g; TOTAL FAT 1.8g; SATURATED FAT 0.2g; SODIUM 18mg; FIBER 2g; BETA-CAROTENE 204ug; VITAMIN C 30mg; CALCIUM 31mg; IRON 0.8mg; FOLATE 50ug; MAGNESIUM 35mg; ZINC 0.7mg; SELENIUM 0.7ug

Great Greens

Serves: 5

2 large bunches kale, tough stems and center ribs removed, chopped
1 bunch Swiss chard, mustard greens, or organic spinach, stems removed, chopped
1 tablespoon Spicy Pecan Vinegar or balsamic vinegar
2 cloves garlic, minced
½ tablespoon VegiZest or other no-salt seasoning blend
1 tablespoon chopped fresh dill or 1 teaspoon dried
2 tablespoons chopped fresh basil or 1 teaspoon dried
Freshly ground black pepper

Steam kale for 10 minutes, add Swiss chard, mustard greens, or spinach, and steam another 5 minutes. Transfer to a medium bowl. In a small bowl, combine vinegar, garlic, VegiZest, dill, basil, and pepper; add to greens, and serve.

PER SERVING: CALORIES 35; PROTEIN 2g; CARBOHYDRATES 7g; TOTAL FAT 0.4g; SODIUM 118mg; FIBER 1.6g; BETA-CAROTENE 5080ug; VITAMIN C 57mg; CALCIUM 85mg; IRON 1.8mg; FOLATE 18ug; MAGNESIUM 52mg; ZINC 0.4mg; SELENIUM 0.8ug

Herbed Barley and Lentils
Serves: 4

½ cup chopped onions
2 cloves garlic, chopped
3 cups coconut water or water blended with a date
1 cup hulled barley*
¼ cup lentils
½ teaspoon Italian seasoning
1 tablespoon chopped fresh dill or ¼ teaspoon dried
1 tablespoon finely chopped fresh chives or 1 teaspoon dried
2 tablespoons finely chopped fresh basil or ½ teaspoon dried

In a large saucepan, sauté onions and garlic in a small amount of coconut water, about 5 minutes or until soft. Add remaining ingredients and bring to a boil. Reduce heat, cover, and simmer until barley and lentils are tender and water is absorbed (about 1 hour).

* Hulled barley is barley that has been minimally processed to remove only the tough inedible outer hull. Do not use pearled barley; it receives additional processing and has a lower nutritional value.

PER SERVING: CALORIES 216; PROTEIN 9g; CARBOHYDRATES 43g; TOTAL FAT 1.2g; SATURATED FAT 0.3g; SODIUM 13mg; FIBER 12.1g; BETA-CAROTENE 70ug; VITAMIN C 3mg; CALCIUM 37mg; IRON 2.7mg; FOLATE 72ug; MAGNESIUM 81mg; ZINC 1.9mg; SELENIUM 18.7ug

Mediterranean Bean and Kale Sauté

Serves: 3

> 2 bunches kale, tough stems and center ribs removed, chopped
> ½ cup chopped dried tomatoes (unsulfured, unsalted)
> 1 medium onion, finely chopped
> 1 cup coarsely chopped shiitake or oyster mushrooms
> 3 cloves garlic, pressed
> 1 tablespoon VegiZest or other no-salt seasoning blend
> 1 cup dried and cooked beans, any type, or canned, no salt added
> 1½ tablespoons Riesling Reserve Vinegar or other sweet vinegar
> 1 tablespoon Dijon mustard
> Red pepper flakes
> Tomato-based pasta sauce

In a small bowl, cover dried tomatoes with water and soak at least 1 hour.

In a large skillet, sauté kale, tomatoes, onion, mushrooms, and garlic in a little water over medium heat for 5 minutes, adding water as needed. Add VegiZest and enough water to keep from scorching. Cover and steam for 10 minutes. Add beans, vinegar, mustard, and red pepper flakes and cook until mushrooms are tender and liquid has cooked out (about 3 minutes). Stir in pasta sauce as desired, and serve.

Note: Top with chicken strips or baked fish filet if you wish.

PER SERVING: CALORIES 168; PROTEIN 9g; CARBOHYDRATES 31g; TOTAL FAT 2.4g; SATURATED FAT 0.3g; SODIUM 146mg; FIBER 6.9g; BETA-CAROTENE 7152ug; VITAMIN C 99mg; CALCIUM 152mg; IRON 2.9mg; FOLATE 86ug; MAGNESIUM 65mg; ZINC 1.2mg; SELENIUM 7.7ug

No-Meat Balls

Serves: 5

½ cup diced onion
4 cloves garlic, roasted* and crushed
¼ cup diced celery
2 tablespoons minced parsley
¼ teaspoon dried sage
2 tablespoons chopped fresh basil or 1 teaspoon dried
1 teaspoon dried oregano
1⅔ cups cooked dried lentils or 1 (15-ounce) can,
    no salt added, drained
¼ cup cooked wild rice
2–3 tablespoons tomato paste
1 tablespoon arrowroot powder or whole wheat flour
2 tablespoons MatoZest or other no-salt seasoning blend
2 tablespoons nutritional yeast
Freshly ground black pepper
2 tablespoons vital wheat gluten flour

In a frying pan, heat 1 tablespoon of water. Add onion and garlic and sauté for 5 minutes. Add celery, parsley, sage, basil, and oregano and sauté another 5 minutes, adding a little more water if needed to prevent sticking.

In a large bowl, combine the sautéed vegetables with the remaining ingredients and mix well. Mash lightly with a potato masher. With wet hands, form 2 tablespoons of the lentil mixture into a smooth ball and place on a very lightly oiled baking sheet. Repeat with the remaining lentil mixture and bake no-meat balls at 350°F for 30 minutes.

Serve with low-salt marinara sauce.

* To roast garlic, break the cloves apart. Leave the papery skins on. Roast at 350°F until soft (about 20 minutes).

PER SERVING: CALORIES 270; PROTEIN 22g; CARBOHYDRATES 45g; TOTAL FAT 1.0g; SATURATED FAT 0.1g; SODIUM 29mg; FIBER 16g; BETA-CAROTENE 494ug; VITAMIN C 8mg; CALCIUM 63mg; IRON 7.4mg; FOLATE 440ug; MAGNESIUM 81mg; ZINC 2.7mg; SELENIUM 8.9ug

Pistachio-Crusted Tempeh with Balsamic Marinade and Shiitake Mushrooms

Serves: 4

8 ounces tempeh, thinly sliced diagonally

Marinade

2 cloves garlic, minced

1 tablespoon chopped fresh cilantro

1 tablespoon chopped fresh basil

Hot pepper flakes

1 cup vegetable broth, no or low salt

2 tablespoons balsamic vinegar

1 teaspoon Bragg Liquid Aminos or low-sodium soy sauce

Crust

1 cup pistachios, shelled

4 tablespoons corn meal

2 tablespoons nutritional yeast

1 teaspoon onion powder

1 teaspoon garlic powder

½ pound shiitake mushrooms, stemmed, thinly sliced

Place tempeh in a saucepan with water to cover, bring to a simmer, and simmer for 10 minutes.

In a bowl, combine all the ingredients for the marinade. Transfer tempeh to the bowl and marinate for at least 1 hour.

Place pistachios in a food processor and process until finely chopped. Add remaining crust ingredients and pulse until thoroughly mixed. Place in large, shallow bowl. Drain tempeh from marinade—reserving the marinade—and dip in crust mixture to coat. Place crusted tempeh and sliced mushrooms side by side on a rimmed baking sheet. Spoon some marinade over mushrooms. Bake at 400°F until mushrooms are soft (13 minutes), turning mushrooms occasionally.

Transfer remaining marinade to a pan and simmer for 2 minutes.
Drizzle tempeh and mushrooms with warm marinade before serving.

PER SERVING: CALORIES 323; PROTEIN 21g; CARBOHYDRATES 21g; TOTAL FAT 20.1g; SATURATED FAT 3g;
SODIUM 68mg; FIBER 5.3g; BETA-CAROTENE 164ug; VITAMIN C 4mg; CALCIUM 117mg; IRON 4.0mg; FOLATE 197ug;
MAGNESIUM 103mg; ZINC 2.3mg; SELENIUM 8.0ug

Ratatouille

Serves: 4

    1 medium onion, thinly sliced
    2 cloves garlic, chopped
    2 large tomatoes, chopped, or 1 (15-ounce) can diced tomatoes,
        no salt added
    1 medium eggplant, cut into 1-inch cubes
    1 medium zucchini, sliced ½ inch thick
    1 medium red bell pepper, cut into 1-inch pieces
    1 teaspoon dried oregano
    1 tablespoon fresh basil, chopped, or 1 teaspoon dried
    Black pepper

In a large deep skillet, sauté the onion in 2 tablespoons water until
softened (about 3 minutes). Add garlic and cook for 1 minute,
adding more water as necessary to keep from scorching. Reduce heat
to moderately low and add tomatoes, eggplant, zucchini, red bell
pepper, oregano, and basil. Add black pepper to taste. Cover and
cook, stirring occasionally, until vegetables are very tender (about 1
hour). Serve warm or at room temperature.

PER SERVING: CALORIES 76; PROTEIN 5g; CARBOHYDRATES 17g; TOTAL FAT 0.6g; SATURATED FAT 0.1g;
SODIUM 16mg; FIBER 7.2g; BETA-CAROTENE 963ug; VITAMIN C 60mg; CALCIUM 43mg; IRON 1.1mg; FOLATE 74ug;
MAGNESIUM 42mg; ZINC 0.6mg; SELENIUM 1ug

Roasted Vegetable Salad with Baked Tofu or Salmon

Serves: 4

    2 red bell peppers, cut into ½-inch pieces
    1 medium eggplant, cut into ½-inch pieces
    1 large yellow squash, cut into ½-inch pieces
    1½ cups butternut squash, peeled and cut into ½-inch pieces
    1 teaspoon olive oil*
    2 tablespoons balsamic vinegar
    3 cloves garlic, minced
    1 teaspoon Bragg Liquid Aminos or low-sodium soy sauce
    ⅛ teaspoon black pepper
    12 cups mixed greens
    Baked Tofu Strips or Baked Salmon (see below)

Lightly coat a large baking pan using a paper towel moistened with olive oil. Place peppers, eggplant, and squash in pan. In a small bowl, mix together olive oil, vinegar, garlic, aminos, and black pepper. Pour over vegetables and toss. Roast at 400°F until tender (18–20 minutes), stirring once. Arrange mixed greens on plates. Remove roasted vegetables from oven and spoon over mixed greens.

Can be served as is or topped with Baked Tofu Strips or Baked Salmon.

Baked Tofu Strips

Cut 1 pound of extra-firm tofu into 1-inch strips. Place on a lightly oiled baking dish and sprinkle with 1 teaspoon each garlic powder and onion powder. Bake at 350°F, turning once, until yellow and firm on the outside and still tender inside (about 30 minutes).

Baked Salmon

Cut 12 ounces of salmon into 4 pieces. Season with garlic powder and black pepper, to taste. Place salmon skin-side down on a nonstick baking sheet. Bake at 450°F until salmon is cooked through (about 12 minutes).

* Only 1 teaspoon of oil (only 10 calories of oil per serving) used here
for 4 servings, so this is acceptable periodically

PER SERVING: (With salmon) CALORIES 155; PROTEIN 12g; CARBOHYDRATES 24g; TOTAL FAT 3.2g;
SATURATED FAT 0.6g; SODIUM 105mg; FIBER 9.4g; BETA-CAROTENE 3364ug; VITAMIN C 107mg; CALCIUM 126mg;
IRON 2.1mg; FOLATE 200ug; MAGNESIUM 81mg; ZINC 1.3mg; SELENIUM 11ug

Saucy Lentil Loaf

Serves: 4

2 large artichokes, rinsed
1 tablespoon water or vegetable broth
½ cup diced onion
2 cloves garlic, minced
3 cups button mushrooms, finely chopped
¼ cup diced celery
2 tablespoons minced parsley
½ teaspoon poultry seasoning
1⅔ cups cooked dried lentils or 1 (15-ounce) can, no salt added,
    rinsed and drained
⅓ cup raw pecans, finely chopped
¼ cup rolled oats
¼ cup tomato paste, plus extra for top of loaf
2 tablespoons arrowroot powder or whole wheat flour
2 tablespoons MatoZest or other no-salt, tomato-based seasoning mix
Freshly ground black pepper

Slice one inch off the top of each artichoke. Cut off the very bottom
of the stems, but keep the stems attached, and slice artichokes in
half lengthwise. Place them in a steamer basket over several inches
of water. Bring water to a boil, cover, and steam for 40 minutes. Set
artichokes aside until cool enough to handle. Scoop out and discard
the fibrous choke from the center of each artichoke. Remove the
hearts and transfer to a bowl. Mash lightly. Scrape off the bottom
third of each leaf and add to mashed hearts. Carefully scrape out the
tender inside of the stems to use as well.

(continued on next page)

In a frying pan, heat water, add onion and garlic, and sauté for 5 minutes. Add mushrooms, cover, and sauté another 4 minutes until mushrooms are tender. Add celery, parsley, and poultry seasoning. Sauté another 5 minutes, adding water if needed to prevent sticking.

Transfer the sautéed vegetables to a bowl and add 1 cup mashed artichokes, lentils, pecans, oats, tomato paste, arrowroot powder, and MatoZest. Sprinkle with black pepper, to taste. Stir well to combine.

Lightly rub a loaf pan with a minimal amount of oil. Fill with the lentil mixture and press down evenly. Spread a ⅛-inch layer of tomato paste over the top. Bake for 1 hour at 350°F. Remove from the oven and let stand at room temperature for 30 minutes before slicing and serving.

PER SERVING: CALORIES 278; PROTEIN 14g; CARBOHYDRATES 43g; TOTAL FAT 7.7g; SATURATED FAT .7g; SODIUM 106mg; FIBER 13.3g; BETA-CAROTENE 936ug; VITAMIN C 20mg; CALCIUM 80mg; IRON 5.6mg; FOLATE 207ug; MAGNESIUM 109mg; ZINC 2.5mg; SELENIUM 12.8ug

Savory Portobello Mushrooms with Chickpeas

Serves: 2

1 large onion, chopped
2 cloves garlic, chopped
3 large Portobello mushroom caps, thinly sliced
⅓ cup red wine or low-sodium vegetable broth
1 large tomato, diced, or 8 halved cherry tomatoes
1⅔ cups cooked garbanzo beans, or 1 (15-ounce) can, no salt added
2 tablespoons water
1 tablespoon chopped fresh mint or 1 teaspoon dried

Water-sauté onion and garlic for a few minutes. Add mushrooms and red wine and cook for 5 more minutes. Add tomatoes, garbanzo beans and water. Cook for another 5–10 minutes. Stir in mint, and serve.

PER SERVING: CALORIES 326; PROTEIN 20g; CARBOHYDRATES 51g; TOTAL FAT 3.9g; SATURATED FAT 0.4g; SODIUM 26mg; FIBER 13.2g; BETA-CAROTENE 828ug; VITAMIN C 31mg; CALCIUM 103mg; IRON 4.8mg; FOLATE 261ug; MAGNESIUM 99mg; ZINC 2.8mg; SELENIUM 12.1ug

Simple Bean Burgers

Serves: 6

- ¼ cup raw sunflower seeds
- 1½ cups cooked dried red or pink beans or 1 (15.5-ounce) can, no salt added, drained
- ½ cup minced onion
- 2 tablespoons ketchup, low sodium
- 1 tablespoon wheat germ or oats
- ½ teaspoon chili powder

Chop the sunflower seeds in a food processor or hand chopper. Mash the beans with a potato masher or in a food processor, transfer to a bowl, and mix with sunflower seed meal. Mix in the remaining ingredients and form into six patties. Lightly coat a baking sheet with olive oil. Place patties on the sheet and bake at 350°F for 25 minutes.

Remove from the oven and let cool (about 10 minutes) until you can pick up each patty and compress it firmly in your hands to re-form the burger. Turn patties over and bake another 10 minutes.

PER SERVING: CALORIES 369; PROTEIN 21g; CARBOHYDRATES 53g; TOTAL FAT 10.2g; SATURATED FAT 1.1g; SODIUM 13mg; FIBER 14.4g; BETA-CAROTENE 187ug; VITAMIN C 8mg; CALCIUM 97mg; IRON 5.5mg; FOLATE 282ug; MAGNESIUM 153mg; ZINC 2.9mg; SELENIUM 13.8ug

Southern-Style Mixed Greens
Serves: 2

1 cup water
1 clove garlic, minced
Black pepper
1⅔ cups cooked dried black-eyed peas* or 1 (15-ounce) can, no
    salt added, rinsed and drained
1 cup chopped yellow pepper
1 cup chopped tomato
⅓ cup chopped fresh parsley
¼ cup chopped red onion
2 tablespoons balsamic vinegar or low-fat dressing
1 (10-ounce) package mixed salad greens

Combine water, garlic, and a dash of black pepper in a large saucepan; bring to a boil. Add black-eyed peas, cover, and simmer over low heat for 10 minutes. Drain.

In a bowl, combine black-eyed peas with yellow peppers, tomato, parsley, onion, and vinegar. Cover and chill for 3 hours or overnight. Serve over salad greens.

* Small white beans can be substituted.

PER SERVING: CALORIES 179; PROTEIN 11g; CARBOHYDRATES 44g; TOTAL FAT 1.3g; SATURATED FAT 0.3g; SODIUM 49mg; FIBER 11.1g; BETA-CAROTENE 1727ug; VITAMIN C 180mg; CALCIUM 279mg; IRON 3.7mg; FOLATE 342ug; MAGNESIUM 119mg; ZINC 2.5mg; SELENIUM 3.9ug

Speedy Vegetable Wrap
Serves: 2

 1 tablespoon low-fat, low-sodium dressing
 2 whole wheat tortillas or whole wheat pitas
 2 cups broccoli slaw mix
 1 large tomato, diced
 ¼ cup red onion, chopped

On each tortilla, spread half of the dressing, the broccoli slaw mix, the tomatoes, and the raw onion. Heat in a toaster oven or microwave just long enough to soften. Roll up tortillas, enclosing filling. If using pitas, stuff with dressing, vegetables, and serve without heating.

Note: Prepackaged broccoli slaw can be purchased in the produce section of many markets. Cole slaw mix, shredded cabbage, or shredded broccoli may also be used.

PER SERVING: CALORIES 184; PROTEIN 11g; CARBOHYDRATES 39g; TOTAL FAT 1.4g; SATURATED FAT 0.1g; SODIUM 155mg; FIBER 8.4g; BETA-CAROTENE 2807ug; VITAMIN C 69mg; CALCIUM 111mg; IRON 1.8mg; FOLATE 59ug; MAGNESIUM 65mg; ZINC 1.3mg; SELENIUM 1.8ug

Swiss Chard with Garlic and Lemon

Serves: 4

> 2 pounds Swiss chard, tough stems and center ribs removed, chopped
> 2 tablespoons water
> 3 cloves garlic, minced
> 1 tablespoon fresh lemon juice
> ¼ teaspoon freshly ground pepper

In a large pan, sauté Swiss chard in water over medium heat until slightly wilted (about 1 minute), adding more water as necessary to keep from scorching. Stir in garlic. Cover and cook until tender (4 minutes), stirring occasionally. Uncover and cook until the liquid evaporates (1–2 minutes). Stir in lemon juice and pepper. Serve.

Note: If you wish, top with choice of healthy dressing.

PER SERVING: CALORIES 37; PROTEIN 3g; CARBOHYDRATES 8g; TOTAL FAT 0.4g; SATURATED FAT 0.1g; SODIUM 363mg; FIBER 2.8g; BETA-CAROTENE 6204ug; VITAMIN C 54mg; CALCIUM 92mg; IRON 3.1mg; FOLATE 24ug; MAGNESIUM 139mg; ZINC 0.6mg; SELENIUM 1.9ug

Thai Vegetable Curry

Serves: 4

4 cloves garlic, finely chopped
2 tablespoons peeled and finely chopped fresh ginger
2 tablespoons chopped fresh mint
2 tablespoons chopped fresh basil
2 tablespoons chopped fresh cilantro
2 cups carrot juice
1 red bell pepper, thinly sliced
1 large eggplant, peeled, if desired, and cut into 1-inch cubes
2 cups green beans, cut in 2-inch lengths
3 cups sliced shiitake mushrooms, stems removed
1 small can bamboo shoots
2 tablespoons VegiZest or other no-salt seasoning
½ teaspoon curry powder
2 cups watercress leaves
3 tablespoons chunky peanut butter, natural, unsalted
1 pound tofu, cut into ¼-inch thick slices, or
  1 pound scallops or shrimp
½ cup light coconut milk
Mint, basil, cilantro, unchopped, for garnish

Add garlic, ginger, mint, basil, cilantro, carrot juice, pepper, eggplant, green beans, mushrooms, bamboo shoots, VegiZest, curry powder, and 1 cup of the watercress to a wok or large skillet. Bring to a boil, cover, and simmer, stirring occasionally, about 10 minutes, until all vegetables are tender. Mix in peanut butter. Add tofu, simmer and toss until hot. Or if you're substituting shrimp or scallops, simmer another 5 minutes until cooked through. Add coconut milk and cook until heated. Top with the remaining 1 cup of watercress and the garnish.

PER SERVING: CALORIES 397; PROTEIN 23g; CARBOHYDRATES 41g; TOTAL FAT 18.5g; SATURATED FAT 7.5g; SODIUM 140mg; FIBER 12.9g; BETA-CAROTENE 12962ug; VITAMIN C 7mg; CALCIUM 262mg; IRON 5.4mg; FOLATE 108ug; MAGNESIUM 175mg; ZINC 2.3mg; SELENIUM 12.1ug

Vegetable Burrito

Serves: 5

1 medium onion, thinly sliced
1 carrot, thinly sliced
1 medium red or green bell pepper, thinly sliced
1 medium zucchini, chopped
¼ teaspoon chili powder
¼ teaspoon cumin
1 medium tomato, chopped
1 avocado, sliced
¼ cup chopped cilantro
5 whole grain tortilla wraps
1 lime

Water-sauté onions, carrot, peppers, and zucchini over medium-high heat until tender (about 6 minutes). Stir in chili powder and cumin.

Spoon sautéed vegetables, tomatoes, 2 slices of avocado, and chopped cilantro onto each tortilla. Add squeeze of lime juice and roll up.

PER SERVING: CALORIES 215; PROTEIN 6g; CARBOHYDRATES 31g; TOTAL FAT 8.9g; SATURATED FAT 1.8g; SODIUM 218mg; FIBER 6.6g; BETA-CAROTENE 600ug; VITAMIN C 58mg; CALCIUM 75mg; IRON 1.8mg; FOLATE 97ug; MAGNESIUM 41mg; ZINC 0.7mg; SELENIUM 7.6ug

Yum Good Beans

Serves: 4

> 3 cups cooked dried kidney beans or 2 (15-ounce) cans, no salt
> added, drained
> 4–6 cloves garlic, minced
> ⅔ cup tomato sauce, no salt added
> 2 teaspoons cumin
> ½ teaspoon cayenne pepper
> 2 tablespoons water

Place all ingredients in a medium pot. Simmer over medium-low heat, stirring often, for 6 minutes. Turn off heat and mash the beans with a masher or fork.

Use these beans to: roll up in several layers of romaine lettuce leaves, stuff into steamed or raw red or green bell peppers, or use as a topping for a salad. Or add a small amount of no-salt, no-oil corn chips with plenty of tomatoes for a taco salad treat.

PER SERVING: CALORIES 193; PROTEIN 13g; CARBOHYDRATES 34g; TOTAL FAT 1g; SATURATED FAT 0.1g; SODIUM 5mg; FIBER 9.6g; BETA-CAROTENE 49ug; VITAMIN C 3mg; CALCIUM 68mg; IRON 3.2mg; FOLATE 185ug; MAGNESIUM 61mg; ZINC 1.5mg; SELENIUM 2.2ug

DESSERTS

Banana Walnut Ice Cream

Serves: 2

    2 ripe bananas, frozen*
    ⅓ cup vanilla soy milk (hemp or almond milk may also be used)
    ½ ounce walnuts

Blend all ingredients together in a high-powered blender or a food processor until smooth and creamy.

* Freeze ripe bananas at least 12 hours in advance. To freeze bananas, peel, cut in thirds and wrap tightly in plastic wrap.

PER SERVING: CALORIES 174; PROTEIN 4g; CARBOHYDRATES 30g; TOTAL FAT 5.9g; SATURATED FAT 0.7g; SODIUM 23mg; FIBER 4.1g; BETA-CAROTENE 178ug; VITAMIN C 10mg; CALCIUM 28mg; IRON 1mg; FOLATE 37ug; MAGNESIUM 53mg; ZINC 0.6mg; SELENIUM 3.5ug

Blueberry Cobbler

Serves: 2

    1 banana, sliced
    1 cup frozen blueberries
    2 tablespoons old-fashioned oats
    1 tablespoon currants
    ¼ teaspoon vanilla
    2 tablespoons chopped raw almonds
    2 tablespoons unsweetened shredded coconut
    ¼ teaspoon cinnamon

Combine banana, berries, oats, currents, and vanilla in a microwave-safe dish. Microwave for 2 minutes. Top with almonds, coconut, and cinnamon and microwave for 1 minute. Serve warm.

PER SERVING: CALORIES 195; PROTEIN 3g; CARBOHYDRATES 35g; TOTAL FAT 6g; SATURATED FAT 1.9g; SODIUM 3mg; FIBER 6.2g; BETA-CAROTENE 39ug; VITAMIN C 8mg; CALCIUM 32mg; IRON 1.4mg; FOLATE 25ug; MAGNESIUM 66mg; ZINC 0.7mg; SELENIUM 4.8ug

Cantaloupe Slush

Serves: 4

1 cantaloupe, seeded, rind removed
2 cups ice
2 Medjool dates or 4 regular dates, pitted

Blend the ingredients together in a high-powered blender until smooth.

Note: The same drink can be made with peaches or nectarines.

PER SERVING: CALORIES 47; PROTEIN 1g; CARBOHYDRATES 11g; TOTAL FAT 0.2g; SATURATED FAT 0.1g; SODIUM 22mg; FIBER 1.2g; BETA-CAROTENE 2787ug; VITAMIN C 51mg; CALCIUM 12mg; IRON 0.3mg; FOLATE 29ug; MAGNESIUM 17mg; ZINC 0.3mg; SELENIUM 0.6ug

Mango Coconut Sorbet

Serves: 4

½ cup unsweetened shredded coconut
2 tablespoons water
¼ teaspoon lemon or lime juice
1 (10-ounce) bag frozen mangos
2 slices dried mango, unsweetened and unsulfured

Blend ingredients in a high-powered blender or a food processor until smooth.

PER SERVING: CALORIES 93; PROTEIN 1g; CARBOHYDRATES 17g; TOTAL FAT 3.6g; SATURATED FAT 3g; SODIUM 4mg; FIBER 2.4g; BETA-CAROTENE 340ug; VITAMIN C 21mg; CALCIUM 10mg; IRON 0.4mg; FOLATE 13ug; MAGNESIUM 11mg; ZINC 0.1mg; SELENIUM 1.5ug

Peach Sorbet

Serves: 4

1 pound frozen peaches
¼ cup soy, almond, or hemp milk
4 dates, pitted

Blend all ingredients in a high-powered blender or food processor until silky smooth.

PER SERVING: CALORIES 138; PROTEIN 2g; CARBOHYDRATES 34g; TOTAL FAT 0.5g; SATURATED FAT 0.1g; SODIUM 15mg; FIBER 2.9g; BETA-CAROTENE 215ug; VITAMIN C 107mg; CALCIUM 12mg; IRON 0.7mg; FOLATE 7ug; MAGNESIUM 13mg; ZINC 0.1mg; SELENIUM 1.4ug

Poached Pears with Raspberry Sauce

Serves: 2

2 pears
1 teaspoon lemon juice
⅔ cup frozen red raspberries, thawed
1 Medjool date, or 2 regular dates, pitted

Peel the pears and leave the stems attached. Drizzle with lemon juice and place in a microwave-safe bowl. Microwave for 4 minutes. Remove pears to individual bowls, reserving cooking liquid.

Blend raspberries and dates in a high-powered blender until smooth. Mix with cooking liquid. Top pears with raspberry sauce, and serve.

PER SERVING: CALORIES 151; PROTEIN 1g; CARBOHYDRATES 40g; TOTAL FAT 0.5g; SATURATED FAT 0.4g; SODIUM 2mg; FIBER 8.6g; BETA-CAROTENE 37ug; VITAMIN C 19mg; CALCIUM 33mg; IRON 0.7mg; FOLATE 22ug; MAGNESIUM 27mg; ZINC 0.4mg; SELENIUM 0.3ug

Strawberry Pineapple Sorbet

Serves: 4

1 (10-ounce) bag frozen strawberries
1 peeled navel orange, or other orange, peeled and seeded
(pre-soaked in water to cover)
4 slices dried pineapple, unsweetened and unsulfured

Blend frozen strawberries, orange, and pineapple with soaking water in a high-powered blender or food processor. Pour into sorbet glasses and top with sliced fresh strawberries, if desired.

PER SERVING: CALORIES 89; PROTEIN 0.4g; CARBOHYDRATES 23g; TOTAL FAT 0.1g; SATURATED FAT 0g; SODIUM 9mg; FIBER 2.6g; BETA-CAROTENE 42ug; VITAMIN C 46mg; CALCIUM 32mg; IRON 0.5mg; FOLATE 22ug; MAGNESIUM 11mg; ZINC 0.1mg; SELENIUM 0.7ug

Take It from Here

I hope this book is the beginning of your journey to lasting health and vibrant living. Many of the suggestions may be different than anything you've tried before, and I acknowledge that this may also require a significant lifestyle change for you and those around you. Change isn't always easy, but it can be immensely rewarding. Watching your diabetes reverse and potentially disappear will be a wonderful experience, especially when you know you created the opportunity by providing your body with the nourishment it needs to do its job.

As you begin your journey to health, remember to plan for your success and address every obstacle as a rewarding challenge to overcome. You can climb Mount Everest and raise your arms in triumph. This is your chance to shine and to glow with pride as you allow your body to glow with health. Recruit your personal support team, and encourage them not just to cheer you on but also to join you. After all, the diet recommendations in this book are also largely the same suggestions I would make to anyone for weight loss, superior health, and longevity. Good luck with your journey. I am privileged to have shared this information with you, and I encourage you to share *your* success story with me. Visit me at www.drfuhrman.com. I will be excited to hear from you.

Acknowledgments

I am grateful for my loving family, my wife Lisa and our three daughters Talia, Jenna, and Cara and our son Sean—and their understanding during my bouts of extreme work hours, especially when completing a book such as this. I appreciate their enthusiasm and support of what I do to help others in need.

I would like to thank my entire team at DrFuhrman.com, whose work is not merely a job but a passion supporting a mission of caring, enabling so many to improve their health. For this book specifically, thanks to Linda Popescu, RD, our food scientist, who assists me with nutritional calculations and scoring, and who helped me choose and tweak the recipes, and Deana Ferreri, Ph.D., our cardiovascular nutrition scientist who aids me in compiling and comparing research articles and their findings.

I would also like to acknowledge the publishing team at HarperOne, whose professionalism and expertise has enabled the broad distribution of this information, especially Gideon Weil, who had a vision and unwavering enthusiasm for the value of my writings to humanity. Additional thanks for the expert production management of Lisa Zuniga, and Melinda Mullin the director of publicity, for her great efforts.

I would like to express my appreciation for the hundreds of progressive medical doctors who have started using my works in their medical practices. I appreciate their openness to learn something not taught in medical school and strong desire to do what is best for their patients. I am extremely moved when they excitedly contact me about the successes they've had with their patients.

Notes

Introduction: A Letter of Hope

1. Dunaief D, Fuhrman J, Dunaief J, Ying G. Glycemic and cardiovascular parameters improved in type 2 diabetes with the high nutrient density (HND) diet. Open Journal of Preventive Medicine 2012; 2: 364–71. doi: 10.4236/ojpm.2012.23053.

Chapter 1: The First Step—Understanding Diabetes

1. Larsson SC, Orsini N, Wolk A. Diabetes mellitus and risk of colorectal cancer: a meta-analysis. J Natl Cancer Inst 2005; 97(22): 1679–87.

2. Economic costs of diabetes in the U.S in 2007. Diabetes Care 2008; 31(3): 596–615.

3. The United States of Diabetes: Challenges and opportunities in the decade ahead. United Health Center for Health Reform and Modernization. Working Paper 5, Nov 2010. http://www.unitedhealthgroup.com/hrm/UNH_Working-Paper5.pdf.

4. Type 2 diabetes—time to change our approach. Lancet 2010; 375(9733): 2193.

5. Zoler ML. Insulin may boost cardiovascular risk in type 2 diabetes patients. Family Practice News May 15, 2001: 6.

6. Madonna R, et al. Insulin enhances vascular cell adhesion molecule-1 expression in human cultured endothelial cells through a pro-atherogenic pathway mediated by p38 mitogen-activated protein-kinase. Diabetologia 2004; 47: 532–6. Taegtmeyer H. Insulin resistance and atherosclerosis, common roots for two common diseases? Circulation 1996; 93: 177.

7. Experts call for further research into the relationship between insulin therapy and cancer. http://www.eurekalert.org/pub_releases/2010-03/w-ecf030210.php. Pollak M, Russell-Jones D. Insulin analogues and cancer risk: cause for concern or cause célèbre? Int J Clin Pract 2010 Apr; 64(5): 628–36.

8. Akbaraly TN, Kivimäki M, Brunner EJ, et al. Association between metabolic syndrome and depressive symptoms in middle-aged adults. Diabetes Care 2009; 32(3): 499–504. Harish K, Dharmalingam M, Himanshu M. Study protocol: insulin and its role in cancer. BMC Endocr Disord 2007; 7: 10.

9. Laitinen JH, Ahola IE, Sarkkinen ES, et al. Impact of intensified dietary therapy on energy and nutrient intakes and fatty acid composition of serum lipids in patients with recently diagnosed noninsulin-dependent diabetes mellitus. J Am Diet Assoc 1993; 93: 276–83. Eilat-Adar S, Xu J, Zephier E, et al. Adherence to dietary recommendations for saturated fat, fiber, and sodium is low in American Indians and other U.S. adults with diabetes. J Nutr 2008; 138(9): 1699–704.

Chapter 2: Don't Medicate, Eradicate

1. Monash University. Critical link between obesity and diabetes discovered. Science Daily 9 July 2009. 16 August 2009 http://www.sciencedaily.com/releases/2009/07/090708090917.htm.

2. Yang Q, Graham TE, Mody N, et al. Serum retinol binding protein 4 contributes to insulin resistance in obesity and type 2 diabetes. Nature 2005; 436(7049): 356–62.

3. Risérus U, Willett WC, Hu FB. Dietary fats and prevention of type 2 diabetes. Prog Lipid Res 2009; 48(1): 44–51.

4. Williamson DF, Thompson TJ, Thun M, et al. Intentional weight loss and mortality among overweight individuals with diabetes. Diabetes Care 2000; 23(10): 1499–1504.

5. Carter P, Gray LJ, Troughton J, et al. Fruit and vegetable intake and incidence of type 2 diabetes mellitus: systematic review and meta-analysis. BMJ 2010; 341: c4229.

6. Ruige JB, Mertens I, Considine RV, et al. Opposite effects of insulin-like molecules and leptin in coronary heart disease of type 2 diabetes preliminary data. Int J Cardiol 2006 Jul 28; 111(1): 19–25.

7. Zoler ML. Insulin may boost cardiovascular risk in type II diabetes patients. Family Practice News May 15, 2001: 6. Cao W, Ning J, Yang X, Liu Z. Excess exposure to insulin is the primary cause of insulin resistance and its associated atherosclerosis. Curr Mol Pharmacol 2011 Nov; 4(3): 154–66.

8. Harish K, Dharmalingam M, Himanshu M. Study protocol: insulin and its role in cancer. BMC Endocr Disord 2007; 7: 10. Bowker SL, Majumdar SR, Veugelers P, Johnson JA. Increased cancer-related mortality for patients with type 2 diabetes who use sulfonylureas or insulin. Diabetes Care 2006; 29(2): 254–8.

9. Tzoulaki I, Molokhia M, Curcin V, et al. Risk of cardiovascular disease and all cause mortality among patients with type 2 diabetes prescribed oral anti-diabetes drugs: retrospective cohort study using UK general practice research database. BMJ 2009; 339: b4731.

10. Jancin B. Sulphonylureas may cause increased mortality risk. Family Practice News Aug 2012; 34.

11. Schauer PR, Burguera B, Ikramuddin S, et al. Effect of laparoscopic Roux-en Y gastric bypass on type 2 diabetes mellitus. Ann Surg 2003; 238(4): 467–84; discussion 84–85.

12. Harder H, Dinesen B, Astrup A. The effect of a rapid weight loss on lipid profile and glycemic control in obese type 2 diabetic patients. Int J Obes Relat Metab Disord 2004; 28(1): 180–2.

Chapter 3: Standard American Diet Versus a Nutritarian Diet

1. Kanauchi M, Tsujimoto N, Hashimoto T, et al. Advanced glycation end products in non-diabetic patients with coronary artery disease. Diabetes Care 2001; 24(9): 1620–3. Krajcovicova-Kudlackova M, Sebekova K, Schinzel R, et al. Advanced glycation end products and nutrition. Physiol Res 2002; 51: 313–6.

2. Grundy SM, Cleeman JI, Merz CN, Brewer HB Jr, Clark LT, Hunninghake DB, et al. Implications of recent clinical trials for the National Cholesterol Education Program Adult Treatment Panel III guidelines. Circulation 2004; 110: 227–39. American Dietetic Association. Hyperlipidemia Medical Nutrition Therapy Protocol. Chicago: American Dietetic Association, 2001. U.S. Preventive Services Task Force. Behavioral counseling in primary care to promote a healthy diet: recommendations and rationale. Am J Prev Med 2003; 24: 93–100.

3. Atkinson FS, Foster-Powell K, Brand-Miller JC. International tables of glycemic index and glycemic load values: 2008. Diabetes Care 2008 Dec; 31(12): 2281–3.

4. Raben A. Should obese patients be counseled to follow a low-glycaemic index diet? No. Obes Rev 2002 Nov; 3(4): 245–56.

5. Raatz SK, Torkelson CJ, Redmon JB, et al. Reduced glycemic index and glycemic load diets do not increase the effects of energy restriction on weight loss and insulin sensitivity in obese men and women. J Nutr 2005 Oct; 135(10): 2387–91.

6. Jenkins DJ, Kendall CW, Popovich DG, et al. Effect of a very-high-fiber vegetable, fruit, and nut diet on serum lipids and colonic function. Metabolism 2001; 50(4): 494–503.

Chapter 4: Reversing Diabetes Is All About Understanding Hunger

1. Vives-Bauza C, Anand M, Shirazi AK, et al. The age lipid A2E and mitochondrial dysfunction synergistically impair phagocytosis by retinal pigment epithelial cells.

2. Patel, C, Husam G, Shreyas R, et al. Prolonged reactive oxygen species generation and nuclear factor-B activation after a high-fat, high-carbohydrate meal in the obese. J Clin Endocrinology & Metabolism 2007; 92(11): 4476–9.

3. Peairs AT, Rankin JW. Inflammatory response to a high-fat, low-carbohydrate weight loss diet: effect of antioxidants. Obesity 2008; 16(7): 1573–8.

4. Scanlan N. Compromised hepatic detoxification in companion animals and its correction via nutritional supplementation and modified fasting. Altern Med Rev 2001; 6 Suppl: S24–37.

5. Levi F, Schibler U. Circadian rhythms: mechanisms and therapeutic implications. Annu Rev Pharmacol Toxicol 2007; 47: 593–628.

6. Patel C, Husam G, Shreyas R, et al. Prolonged reactive oxygen species generation and nuclear factor-b activation after a high-fat, high-carbohydrate meal in the obese. J Clin Endocrinology & Metabolism 2007; 92(11): 4476–9.

7. Peairs AT, Rankin JW. Inflammatory response to a high-fat, low-carbohydrate weight loss diet: effect of antioxidants. Obesity 2008; 16(7): 1573–8.

8. Fuhrman J, Sarter B, Glaser D, Acocella S. Changing perceptions of hunger on a high nutrient density diet. Nutrition Journal 2010; 9:51.

Chapter 5: High-Protein, Low-Carb Counterattack

1. Best TH, Franz DN, Gilbert DL, et al. Cardiac complications in pediatric patients on the ketogenic diet. Neurology 2000; 54(12): 2328–30.

2. Best TH, Franz DN, Gilbert DL, et al. Cardiac complications in pediatric patients on the ketogenic diet. Neurology 2000; 54(12): 2328–30.

3. Stevens A, Robinson DP, Turpin J, et al. Sudden cardiac death of an adolescent during (Atkins) dieting. Southern Medical Journal 2002; 95: 1047.

4. Newgard CB, An J, Bain JR, et al. A branched-chain amino acid-related metabolic signature that differentiates obese and lean humans and contributes to insulin resistance. Cell Metabolism 2009; 9(4): 311–26.

5. Sluijs I, Beulens JWJ, Van Der A DL, et al. Dietary intake of total animal and vegetable protein and risk of type 2 diabetes in the European prospective investigation into cancer and nutrition (EPIC)-NL study. Diabetes Care 2010; 33: 43–48.

6. Tonstad S, Butler T, Yan R, Fraser GE. Type of vegetarian diet, body weight, and prevalence of type 2 diabetes. Diabetes Care 2009; 32: 791–6.

7. Jenkins DJ, Kendall CW, Popovich DG, et al. Effect of a very-high-fiber vegetable, fruit, and nut diet on serum lipids and colonic function. Metabolism 2001 Apr; 50(4): 494–503.

8. Fleming RM. The effect of high-protein diets on coronary blood flow. Angiology 2000; 51(10): 817–26.

9. Lagiou P, Sandin S, Lof M, et al., Low carbohydrate–high protein diet and incidence of cardiovascular diseases in Swedish women: prospective cohort study. BMJ 2012; 344: e406.

10. Knight EL, Stampfer, MJ, Hankinson SE, et al. The impact of protein on renal function decline in women with normal renal function or mild renal insufficiency. Ann Int Med 2003; 138: 460–7.

11. Atkins diet raises concerns. Cortland Forum 2004 (April): 22.

12. American Kidney Fund press release, April 25, 2002, http://www.kidney fund.org/AboutAKF/newsroom–020425.htm.

13. Kaushik M, Mozaffarian D, Spiegelman D, et al. Long-chain omega-3 fatty acids, fish intake, and the risk of type 2 diabetes mellitus. Am J Clin Nutr 2009; 90: 613–20.

14. Qi L, Van Dam RN, Rexrodck, Hu FB. Heme iron from diet as a risk factor for coronary heart disease in woman with type 2 diabetes. Diabetes Care 2007; 30(1): 101–6.

15. Hu FB. Associations of dietary protein with disease and mortality in a prospective study of postmenopausal women. Am J Epidemiol 2005; 161(3): 239–49. Kelemen LE, Kushi LH, Jacobs DR, Cerhan JR. Plant-based foods and prevention of cardiovascular disease: an overview. Am J Clin Nutr 2003; 78(3 Suppl): 544S–551S.

16. Lutsey PL, Steffen LM, Stevens J. Dietary intake and the development of the metabolic syndrome. the atherosclerosis risk in communities study. Circulation 2008; 117: 754–61.

17. Gardner CD, Coulston A, Chatterjee L, et al. The effect of a plant-based diet on plasma lipids in hypercholesterolemic adults: a randomized trial. Ann Intern Med 2005; 142(9): 725–33. Tucker KL, Hallfrisch J, Qiao N, et al. The combination of high fruit and vegetable and low saturated fat intakes is more protective against mortality in aging men than is either alone: the Baltimore Longitudinal Study of Aging. J Nutr 2005; 135(3): 556–61. Campbell TC, Parpia B, Chen J. Diet, lifestyle, and the etiology of coronary artery disease: the Cornell China study. Am J Cardiol 1998 Nov 26; 82(10B): 18T–21T.

18. Diousse L, Gaziano JM, Buring JC, et al. Egg consumption and risk of type 2 diabetes in men and women. Diabetes Care 2008; 10: 2337/1271.

19. Netteton JA, Steffen LM, Loehr LF, et al. Incident heart failure is associated with lower whole-grain intake and greater high fat dairy and egg intake in the atheroscerosis risk in communities (ARIC) study. J Am Dietetic Assoc 2008; 108(11): 1881–7.

20. Hu FB, Stampfer MJ, Rimm EB, et al. A prospective study of egg consumption and risk of cardiovascular disease in men and women. JAMA 1999; 281: 1387–94.

21. Trichopoulou A, Psaltopoulou T, Orfanos P, et al. Diet and physical activity in relation to overall mortality amongst adult diabetics in a general population cohort. J Intern Med 2006; 259: 583–91.

22. Spence JD, Eliasziw M, DiCicco M, et al. Carotid plaque area: a tool for targeting and evaluating vascular preventive therapy. Stroke 2002; 33: 2916–22.

23. Djousse L, Gaziano JM, Buring JE, et al. Egg consumption and risk of type 2 diabetes in men and women. Diabetes Care 2009; 32: 295–300. Richman EL, Kenfield SA, Stampfer MJ, et al. Egg, red meat, and poultry intake and risk of lethal prostate cancer in the prostate-specific antigen-era: incidence and survival. Cancer Prev Res (Phila) 2011; 4: 2110–21.

24. Helman AD, Darnton-Hill I, Craig WJ, et al. Iron status of vegetarians. Am J Clin Nutr 1994; 59 (suppl 5): 1203S–1212S.

25. Rose W. The amino acid requirements of adult man. Nutritional Abstracts and Reviews 1957; 27: 631.

26. Hardage M. Nutritional studies of vegetarians. Journal of the American Dietetic Association 1966; 48: 25.

27. Bartke A. Minireview: role of the growth hormone/insulin-like growth factor system in mammalian aging. Endocrinology 2005; 146(9): 3718–23.

28. Laron Z. The GH-IGF1 axis and longevity: the paradigm of IGF1 deficiency. Hormones (Athens) 2008; 7(1): 24–27. Berryman DE, et al. Role of the GH/IGF-1 axis in lifespan and healthspan: lessons from animal models. Growth Horm IGF Res 2008; 18(6): 455–71. Van Bunderen CC, et al. The association of serum insulin-like growth factor-i with mortality, cardiovascular disease, and cancer in the elderly: a population-based study. J Clin Endocrinol Metab 2010.

29. Kraemer WJ, Ratamess NA. Hormonal responses and adaptations to resistance exercise and training. Sports Med 2005; 35(4): 339–61. Allen NE, et al. Lifestyle determinants of serum insulin-like growth-factor-I (IGF-I), C-peptide and hormone binding protein levels in British women. Cancer Causes Control 2003; 14(1): 65–74.

30. Gualberto A, Pollak M. Emerging role of insulin-like growth factor receptor inhibitors in oncology: early clinical trial results and future directions. Oncogene 2009; 28(34): 3009–21.

31. Salvioli S, et al. Why do centenarians escape or postpone cancer? the role of IGF-1, inflammation and p53. Cancer Immunol Immunother 2009; 58(12): 1909–17. Chitnis MM, et al. The type 1 insulin-like growth factor receptor pathway. Clin Cancer Res 2008; 14(20): 6364–70.

32. Rinaldi S, et al. IGF-I, IGFBP-3 and breast cancer risk in women: the European prospective investigation into cancer and nutrition (EPIC). Endocr Relat Cancer 2006; 13(2): 593–605.

33. Hankinson SE, et al. Circulating concentrations of insulin-like growth factor-I and risk of breast cancer. Lancet 1998; 351(9113): 1393–6.

34. Lann D, LeRoith D. The role of endocrine insulin-like growth factor-I and insulin in breast cancer. J Mammary Gland Biol Neoplasia 2008; 13(4): 371–9. Allen NE, et al. A prospective study of serum insulin-like growth factor-I (IGF-I), IGF-II, IGF-binding protein-3 and breast cancer risk. Br J Cancer 2005; 92(7): 1283–7. Fletcher O, et al. Polymorphisms and circulating levels in the insulin-like growth factor system and risk of breast cancer: a systematic review. Cancer Epidemiol Biomarkers Prev 2005; 14(1): 2–19. Renehan AG, et al. Insulin-like growth factor (IGF)-I, IGF binding protein-3, and cancer risk: systematic review and meta-regression analysis. Lancet 2004; 363(9418): 1346–53. Shi R, et al. IGF-I and breast cancer: a meta-analysis. Int J Cancer 2004; 111(3): 418–23. Sugumar A, et al. Insulin-like growth factor (IGF)-I and IGF-binding protein 3 and the risk of premenopausal breast cancer: a meta-analysis of literature. Int J Cancer 2004; 111(2): 293–7. Baglietto L, et al. Circulating insulin-like growth factor-I and binding protein-3 and the risk of breast cancer. Cancer Epidemiol Biomarkers Prev 2007; 16(4): 763–8.

35. Davies M, et al. The insulin-like growth factor system and colorectal cancer: clinical and experimental evidence. Int J Colorectal Dis 2006; 21(3): 201–8. Sandhu MS, Dunger DB, Giovannucci EL. Insulin, insulin-like growth factor-I (IGF-I), IGF binding proteins, their biologic interactions, and colorectal cancer. J Natl Cancer Inst 2002; 94(13): 972–80.

36. Rowlands MA, et al. Circulating insulin-like growth factor peptides and prostate cancer risk: a systematic review and meta-analysis. Int J Cancer 2009; 124(10): 2416–29.

37. Hirano S, et al. Clinical implications of insulin-like growth factors through the presence of their binding proteins and receptors expressed in gynecological cancers. Eur J Gynaecol Oncol 2004; 25(2): 187–91. Menu E, et al. The role of the insulin-like growth factor 1 receptor axis in multiple myeloma. Arch Physiol Biochem 2009; 115(2): 49–57. Rikhof B, et al. The insulin-like

growth factor system and sarcomas. J Pathol 2009; 217(4): 469–82. Parker AS, et al. High expression levels of insulin-like growth factor-I receptor predict poor survival among women with clear-cell renal cell carcinomas. Hum Pathol 2002; 33(8): 801–5.

38. Giovannucci E, et al. Nutritional predictors of insulin-like growth factor I and their relationships to cancer in men. Cancer Epidemiol Biomarkers Prev 2003; 12(2): 84–9. Thissen JP, Ketelslegers JM, Underwood LE. Nutritional regulation of the insulin-like growth factors. Endocr Rev 1994; 15(1): 80–101.

39. Qin LQ, He K, Xu JY. Milk consumption and circulating insulin-like growth factor-I level: a systematic literature review. Int J Food Sci Nutr 2009; 60 Suppl 7: 330–40. Lu FR, Shen L, Qin Y, et al. Clinical observation on trigonella foenum-graecum L. total saponins in combination with sulfonylureas in the treatment of type 2 diabetes mellitus. Chin J Integr Med 2008, 14:56–60.

40. Fontana L, et al. Long-term effects of calorie or protein restriction on serum IGF-1 and IGFBP-3 concentration in humans. Aging Cell 2008; 7(5): 681–7. Mudra M, Ercan-Fang N, Zhong L, et al. Influence of mulberry leaf extract on the blood glucose and breath hydrogen response to ingestion of 75 g sucrose by type 2 diabetic and control subjects. Diabetes Care 2007, 30:1272–74. Judy WV, Hari SP, Stogsdill WW, et al. Antidiabetic activity of a standardized extract (Glucosol) from Lagerstroemia speciosa leaves in Type II diabetics. A dose-dependence study. J Ethnopharmacol 2003, 87:115–117.

41. Kaaks R. Nutrition, insulin, IGF-1 metabolism and cancer risk: a summary of epidemiological evidence. Novartis Found Symp 2004; 262: 247–60; discussion 260–68. McCarty MF. Vegan proteins may reduce risk of cancer, obesity, and cardiovascular disease by promoting increased glucagon activity. Med Hypotheses 1999; 53(6): 459–85. Cannata D, et al. Type 2 diabetes and cancer: what is the connection? Mt Sinai J Med 2010; 77(2): 197–213. Venkateswaran V, et al. Association of diet-induced hyperinsulinemia with accelerated growth of prostate cancer (LNCaP) xenografts. J Natl Cancer Inst 2007; 99(23): 1793–800.

Chapter 6: The Phenomenal Fiber in Beans

1. Omiea I, Lazcano-Ponce E, Sanchez-Zamorano LM, et al. Carbohydrates and the risk of breast cancer among Mexican women. Cancer Epidemiol Biomarkers Prev 2004; 13: 1283–9.

2. Finley JW, Burrell JB, Reeves PG, et al. Pinto bean consumption changes SCFA profiles in fecal fermentations, bacterial populations of the lower bowel, and lipid profiles in blood of humans. J Nutr 2007; 137(11): 2391–8.

3. Robertson MD, Currie JM, Morgan LM, et al. Prior short-term consumption of resistant starch enhances postprandial insulin sensitivity in healthy subjects. Diabetologia 2003; 46(5): 659–65.

4. Higgins JA, Higbee DR, Donahoo WT, et al. Resistant starch consumption promotes lipid oxidation. Nutrition & Metabolism 2004; 1:8doi:10.1186/1743-7075-1-8.

5. Carter P, Gray LJ, Troughton J, et al. Fruit and vegetable intake and incidence of type 2 diabetes mellitus: systematic review and meta-analysis. BMJ 2010; 341: c4229.

6. Behall KM, Howe JC. Effect of long-term consumption of amylose vs amylopectin starch on metabolic variables in human subjects. American Journal of Clinical Nutrition 1995; 61: 334–40. Jenkins DJ, Vuksan V, Kendall CW, et al. Physiological effects of resistant starches on fecal bulk, short chain fatty acids, blood lipids and glycemic index. Journal of the American College of Nutrition 1998; 17: 609–16.

7. Lanza E, Hartman TJ, Albert PS, et al. High dry bean intake and reduced risk of advanced colorectal adenoma recurrence among participants in the polyp prevention trial. J Nutr 2006; 136(7): 1896–903.

8. Singh PN, Fraser GE. Dietary risk factors for colon cancer in a low-risk population. Am J Epidem 1988; 148: 761–74. Aune D, De Stefani E, Ronco A, et al. Legume intake and the risk of cancer: a multisite case-control study in Uruguay. Cancer Causes Control 2009; 20(9): 1605–15. Agurs-Collins T, Smoot D, Afful J, et al. Legume intake and reduced colorectal adenoma risk in African-Americans. J Natl Black Nurses Assoc 2006; 17(2): 6–12. Lanza E, Hartman TJ, Albert PS, et al. High dry bean intake and reduced risk of advanced colorectal adenoma recurrence among participants in the polyp prevention trial. J Nutr 2006; 136(7): 1896–903.

9. Blackberry I, Kouris-Blazos A, Wahlqvist ML, et al. Legumes: the most important dietary predictor of survival in older people of different ethnicities. Asia Pac J Clin Nutr 2004; 13(Suppl): S126.

10. Wu AH, Yu MC, Tseng CC, Pike MC. Epidemiology of soy exposures and breast cancer risk. Br J Cancer 2008; 98(1): 9–14.

11. Bednar GE, Patil AR, Murray SM, Grieshop CM, Merchen NR, Fahey GC Jr. Starch and fiber fractions in selected food and feed ingredients affect their small intestinal digestibility and fermentability and their large bowel fermentability in vitro in a canine model. J Nutr 2001 Feb; 131(2): 276–86. Muir JG, O'Dea K. Validation of an in vitro assay for predicting the amount of starch that escapes digestion in the small intestine of humans. Am J Clin Nutr 1993 Apr; 57(4): 540–6.

12. Sluijs I, et al. Carbohydrate quantity and quality and risk of type 2 diabetes in the European Prospective Investigation into Cancer and Nutrition–Netherlands (EPIC–NL) study. Am J Clin Nutr 2010; 92(4): 905–11. Barclay AW, et al. Glycemic index, glycemic load, and chronic disease risk—a meta-analysis of observational studies. Am J Clin Nutr 2008; 87(3): 627–37. Gnagnarella P, et al. Glycemic index, glycemic load, and cancer risk: a meta-analysis. Am J Clin Nutr 2008; 87(6): 1793–801. Sieri S, et al. Dietary glycemic load and index and risk of coronary heart disease in a large Italian cohort: the EPICOR study. Arch Intern Med 2010; 170(7): 640–7.

13. Buyken AE, Toeller M, Heitkamp G, et al. Glycemic index in the diet of European outpatients with type 1 diabetes: relations to glycated hemoglobin and serum lipids. Am J Clin Nutr 2001; 73(3): 574–81.

14. Halton T, Willett WC, Liu S, et al. Potato and french fry consumption and risk of type 2 diabetes in women. Am J Clin Nutr 2006; 83(2): 284–90.

15. Hodge AM, et al. Dietary patterns and diabetes incidence in the Melbourne collaborative cohort study. Am J Epidemiol 2007; 165(6): 603–10. Van Dam, RM, et al. Dietary patterns and risk for type 2 diabetes mellitus in U.S. men. Ann Intern Med 2002; 136(3): 201–9.

16. Atkinson FS, Foster-Powell K, Brand-Miller JC. International tables of glycemic index and glycemic load values 2008. Diabetes Care 2008 Dec; 31(12): 2281–3. Foster-Powell K, Holt SH, Brand-Miller JC. International table of glycemic index and glycemic load values: 2002. Am J Clin Nutr 2002 Jul; 76(1): 5–56.

Chapter 7: The Truth About Fat

1. Hu FB, Willett WC. Optimal diets for prevention of coronary heart disease. JAMA 2002; 288(20): 2569–78. Sabaté J. Nut consumption, vegetarian diets, ischemic heart disease risk, and all-cause mortality: evidence from epidemiologic studies. American Journal of Clinical Nutrition, Vol. 70, No. 3, 500S–503S, September 1999.

2. Hu FB, Stampfer MJ. Nut consumption and risk of coronary heart disease: a review of epidemiologic evidence. Curr Atheroscler Rep 1999 Nov; 1(3): 204–209.

3. Mukuddem-Petersen J, Oosthuizen W, Jerling JC. A systematic review of the effects of nuts on blood lipid profiles in humans. J Nutr 2005; 135(9): 2082–9.

4. Lamarche B, Desroche S, Jenkins DJ, et al. Combined effects of a dietary portfolio of plant sterols, vegetable protein, viscous fiber and almonds on LDL particle size. Br J Nutr 2004; 92(4): 654–63.

5. Cerdá B, Tomás-Barberán FA, Espín JC. Metabolism of antioxidant and chemopreventive ellagitannins from strawberries, raspberries, walnuts, and oak-

aged wine in humans: identification of biomarkers and individual variability. J Agric Food Chem 2005; 53(2): 227–35.

6. Ros E, Naatez I, Parez-Heras A, et al. A walnut diet improves endothelial function in hypercholesterolemic subjects: a randomized crossover trial. Circulation 2004; 109(13): 1609–14.

7. Ellsworth JL, Kushi LH, Folsom AR. Frequent nut intake and risk of death from coronary heart disease and all causes in postmenopausal women: the Iowa Women's Health Study. Nutr Metab Cardiovasc Dis 2001 Dec; 11(6): 372–7. Li TY, Brennan AM, Wedick NM, et al. Regular consumption of nuts is associated with a lower risk of cardiovascular disease in women with type 2 diabetes. J Nutr 2009; 139(7): 1333–8.

8. Albert CM, Gaziano JM, Willett WC, Manson JE. Nut consumption and decreased risk of sudden cardiac death in the Physicians' Health Study. Arch Intern Med 2002 Jun 24; 162(12): 1382–7. Fraser GE, Sabaté J, Beeson WL, Strahan TM. A possible protective effect of nut consumption on risk of coronary heart disease. The Adventist Health Study. Arch Intern Med 1992 Jul; 152(7): 1416–24. Hu FB, Stampfer MJ, Manson JE, et al. Frequent nut consumption and risk of coronary heart disease in women: prospective cohort study. BMJ 1998 Nov 14; 317(7169): 1341–5. Brown L, Rosner B, Willett WC, Sacks F. Nut consumption and risk of recurrent coronary heart disease. FASEB J 1999; 13: A538. Ellsworth JL, Kushi LH, Folsom AR. Frequent nut intake and risk of death from coronary heart disease and all causes in postmenopausal women: the Iowa Women's Health Study. Nutr Metab Cardiovasc Dis 2001 Dec; 11(6): 372–7.

9. Zelman KM. It's full of fat and helps you lose weight. WebMD. http://www.webmd.com/diet/features/its-full-of-fat-and-helps-you-lose-weight.

10. Yuen AW, Sander JW. Is omega-3 fatty acid deficiency a factor contributing to refractory seizures and SUDEP? a hypothesis. Seizure 2004 Mar; 13(2): 104–7.

11. Coates AM, Howe PR. Edible nuts and metabolic health. Curr Opin Lipidol 2007; 18(1): 25–30. Segura R, Javierre C, Lizarraga MA, Ros E. Other relevant components of nuts: phytosterols, folate and minerals. Br J Nutr 2006; 96 Suppl 2: S36–44.

12. Rajaram S, Sabat AJ. Nuts, body weight and insulin resistance. Br J Nutr 2006; 96 Suppl 2: S79–86. Sabat AJ. Nut consumption and body weight. Am J Clin Nutr 2003; 78(3 Suppl): 647S–650S. Bes-Rastrollo M, Sabat AJ, Gamez-Garcia E, et al. Nut consumption and weight gain in a Mediterranean cohort: the SUN study. Obesity 2007; 15(1): 107–16. Garcia-Lorda P, Megias Rangil I, Salas-Salvada J. Nut consumption, body weight and insulin resistance. Eur J Clin Nutr 2003; 57 Suppl 1: S8–11. Megas-Rangil I, Garcia-Lorda P, Torres-Moreno M, et al. Nutrient content and health effects of nuts. Arch Latinoam Nutr 2004; 54(2 Suppl 1): 83–6.

13. Lovejoy JC. The impact of nuts on diabetes and diabetes risk. Curr Diab Rep 2005; 5(5): 379–84. Jiang R, Manson JE, Stampfer MJ, Liu S, Willett WC, Hu FB. Nut and peanut butter consumption and risk of type 2 diabetes in women. JAMA 2002; 288(20): 2554–60.

14. Barnard ND, Cohen J, Jenkins DJ, et al. A low-fat vegan diet improves glycemic control and cardiovascular risk factors in a randomized clinical trial in individuals with type 2 diabetes. Diabetes Care 2006; 29(8): 1777–83. Ford ES, Mokdad AH. Fruit and vegetable consumption and diabetes mellitus incidence among U.S. adults. Prev Med 2001; 32(1): 33–39. Montonen J, Knekt P, Harkanen T, et al. Dietary patterns and the incidence of type 2 diabetes. Am J Epidem 2004; 161(3): 219–27.

15. Barnard ND, Scialli AR, Bertron P, et al. Effectiveness of a low-fat vegetarian diet in altering serum lipids in healthy premenopausal women. Am J Cardiol 2000 Apr 15; 85(8): 969–72.

16. Jenkins DJ, Kendall CW, Popovich DG, et al. Effect of a very-high-fiber vegetable, fruit, and nut diet on serum lipids and colonic function. Metabolism 2001 Apr; 50(4): 494–503.

17. Tsai CJ, Leitzmann MF, Hu FB, Willett WC, Giovannucci EL. Frequent nut consumption and decreased risk of cholecystectomy in women. Am J Clin Nutr 2004; 80(1): 76–81.

18. Tsai CJ, Leitzmann MF, Hu FB, et al. A prospective cohort study of nut consumption and the risk of gallstone disease in men. Am J Epid 2004; 160(10): 961–8.

19. Brown MJ, Ferruzzi MG, Nguyen ML, et al. Carotenoid bioavailability is higher from salads ingested with full-fat than with fat-reduced salad dressings as measured with electrochemical detection. Am J Clin Nutr 2004; 80(2): 396–403.

20. Fraser GE, Shavlik DJ. Ten years of life: is it a matter of choice? Arch Intern Med 2001; 161(13): 1645–52.

21. Novotny JA, Gebauer SK, Baer DJ. Discrepancy between the Atwater factor predicted and empirically measured energy values of almonds in human diets. Am J Clin Nutr 2012; 96(2): 296–301.

Chapter 8: The Nutritarian Diet in Action

1. Stengle J. Diabetes: some beat it, but are they cured? Seattle Times, April 20, 2009. http://seattletimes.com/html/health/2009084495_apmedbeatingdiabetes .html.

2. Link LB, Potter JD. Raw versus cooked vegetables and cancer risk. Cancer Epidemiol Biomarkers Prev 2004; 13(9): 1422–35.

3. Miller AB. Nutritional aspects of human carcinogenesis. IARC Sci Publ 1982; (39): 177–92.

4. Carter P, Gray LJ, Troghton J, et al. Fruit and vegetable intake and incidence of type 2 diabetes mellitus: systematic review and meta-analysis. BMJ 2010; 341: c4229.

5. Liu X, Lv K. Cruciferous vegetables intake is inversely associated with risk of breast cancer: a meta-analysis. Breast 2012 Aug 6. Liu B, Mao Q, Cao M, Xie L. Cruciferous vegetables intake and risk of prostate cancer: a meta-analysis. Int J Urol 2012 Feb; 19(20: 134–41. doi: 10.1111/j.1442-2042.2011.02906.x. World Cancer Research Fund/American Institute for Cancer Research. Food, nutrition, physical activity, and the prevention of cancer: a global perspective. Washington, DC: AICR, 2007. Zhang CX, Ho SC, Chen YM, Fu, JH, Cheng SZ, Lin FY. Greater vegetable and fruit intake is associated with a lower risk of breast cancer among Chinese women. Int J Cancer 2009 Jul 1; 125(1): 181–8.

6. Michaud DS, Spiegelman D, Clinton SK. Fruit and vegetable intake and incidence of bladder cancer in a male prospective cohort. J Natl Cancer Inst 1999; 91(7): 605–13.

7. Gamet-Payrastre L, Lumeau S, Cassar G. Sulforaphane, a naturally occurring isothiocyanate, induces cell cycle arrest and apoptosis in HT29 human colon cancer cells. Cancer Research 2000; 60(5): 1426–33.

8. Cohen JH, Kristal AR, Stanford JL. Fruit and vegetable intake and prostate cancer risk. J Nat Can Inst 2000; 92(1): 61–68.

9. Zakkar M, Van der Heiden K, Anh Luong Le, et al. Activation of Nrf2 in endothelial cells protects arteries from exhibiting a proinflammatory state. Arteriosclerosis Thrombosis and Vascular Biology. Published online before print September 3, 2009, doi: 10.1161/ATVBAHA.109.193375.

10. Seddon JM, Ajani UA, Sperduto RD, et al. Dietary carotenoids, vitamins A, C, and E, and advanced age-related macular degeneration. JAMA 1994; 272: 1413–20.

11. Dwyer JH, Navab M, Dwyer KM, et al. Oxygenated carotenoid lutein and progression of early atherosclerosis: the Los Angeles Atherosclerosis Study. Circulation 2001; 103(24): 2922–7.

12. Bellisle F. Impact of the daily meal pattern on energy balance. Scandinavian Journal of Nutrition 2004; 48: 114–18. Marmonier C, Chapelot D, Fantino M, Louis-Sylvestre J. Snack consumed in a non hungry state has poor satiating efficiency: influence of snack composition on substrate utilization and hunger. American Journal of Clinical Nutrition 2002; 76: 518–28. Favero A, Franceschi S, La Vecchia C, et al. Meal frequency and coffee intake in colon cancer. Nutr Cancer 1998; 30(3): 182–5. Stote KS, Baer DJ, Spears K, et al. GKA controlled

trial of reduced meal frequency without caloric restriction in healthy, normal-weight, middle-aged adults. Am J Clin Nutr 2007; 85(4): 981–8. Bertéus FH, Torgerson JS, Sjöström L, Lindroos AK. Snacking frequency in relation to energy intake and food choices in obese men and women compared to a reference population. International Journal of Obesity 2005; 29(6): 711–9.

13. Figueiredo JC, Grau MV, Haile RW, et al. Folic acid and risk of prostate cancer: results from a randomized clinical trial. J Natl Cancer Inst 2009; 101(6): 432–5. Ebbing M, Bonaa KH, Arnesen E, et al. Cancer incidence and mortality after treatment with folic acid and vitamin B$_{12}$. JAMA 2009; 302(19): 2119–26. Charles D, Ness AR, Campbell D, et al. Taking folate in pregnancy and risk of maternal breast cancer. BMJ 2004; 329: 1375–6. Fife J, Raniga S, Hider PN, Frizelle FA. Folic acid supplementation and colorectal cancer risk; a meta-analysis. Colorectal Dis 2011; 13(2): 132–7. Stolzenberg-Solomon RZ, Chang S, Leitzmann MF, et al. Folate intake, alcohol use, and postmenopausal breast cancer risk in the prostate, lung, colorectal, and ovarian cancer screening trial. Am J Clin Nutr 2006; 83(4): 895–904. Yi K. Does a high folate intake increase the risk of breast cancer? Nutr Rev 2006; 64(10PT1): 468–75. Cole B, Baron J, Sandler R, et al. Folic acid for the prevention of colorectal adenomas. JAMA 2007; 297(21): 2351–9. Smith AD, Kim Y, et al. Is folic acid good for everyone? Am J Clin Nutr 2008; 87(3): 517. Kim Y. Role of folate in colon cancer development and progression. J Nutr 2003 133(11 Suppl): 3731S–3739S. Guelpen BV, Hultdin J, Johansson I, et al. Low folate levels may protect against colorectal cancer. Gut 2006; 55: 1461–6.

14. Mayne ST. Beta-carotene, carotenoids, and disease prevention in humans. FASEB 1996; 10(7): 690–701. Goodman GE. Prevention of lung cancer. Current Opinion in Oncology 1998; 10(2): 122–6. Kolata G. Studies find beta carotene, taken by millions, can't forestall cancer or heart disease. New York Times, Jan 19, 1996. Omenn GS, Goodman GE, Thornquist MD, et al. Effects of a combination of beta carotene and vitamin a on lung cancer and cardiovascular disease. New England Journal of Medicine 1996; 334(18); 1150–5. Hennekens CH, Buring JE, Manson JE, et al. Lack of effect of long-term supplementation with beta carotene on the incidence of malignant neoplasms and cardiovascular disease. New England Journal of Medicine 1996; 334(18): 1145–9. Albanes D, Heinonen OP, Taylor PR, et al. Alpha-tocopherol and beta-carotene supplements and lung cancer incidence in the alpha-tocopherol, beta-carotene cancer prevention study: effects of base-line characteristics and study compliance. Journal of the National Cancer Institute 1996; 88(21): 1560–70. Rapola JM, Virtamo J, Ripatti S, et al. Randomized trial of alpha-tocopherol and beta-carotene supplements on incidence of major coronary events in men with previous myocardial infarction. Lancet 1997; 349(9067): 1715–20.

15. Omenn GS, Goodman GE, Thornquist MD, et al. Effects of a combination of beta carotene and vitamin A on lung cancer and cardiovascular disease. New England Journal of Medicine 1996; 334(18): 1145–9.

16. Lawson KA, Wright ME, Subar A, et al. Multivitamin use and risk of prostate cancer in the National Institutes of Health-AARP Diet and Health Study. J Natl Cancer Inst 2007 May 16; 99(10): 754–64.

17. Bjelakovic G, Nikolava D, Gluud LL, et al. Antioxidant supplements for prevention of mortality in healthy participants and patients with various diseases. Cochrane Database Syst Rev 2008 Apr 16; (2): CD007176.

18. Hennekens CH, Buring JE, Manson JE, et al. Lack of effect of long-term supplementation with beta carotene on the incidence of malignant neoplasms and cardiovascular disease. New England Journal of Medicine 1996; 334(18): 1145–9. Albanes D, Heinonen OP, Taylor PR, et al. Alpha-tocopherol and beta-carotene supplements and lung cancer incidence in the alpha-tocopherol, beta-carotene cancer prevention study: effects of base-line characteristics and study compliance. Journal of the National Cancer Institute 1996; 88(21): 1156–1570. Rapola JM, Viramo J, Ripatti S, et al. Randomized trial of alpha-tocopherol and beta-carotene supplements on incidence of major coronary events in men with previous myocardial infarction. Lancet 1997; 349(9067): 1715–20.

19. Melhus H, Michaelson K, Kindmark A, et al. Excessive dietary intake of vitamin A is associated with reduced bone mineral density and increased risk of hip fracture. Ann Intern Med 1998; 129(10): 770–8.

20. Bjelakovic G, Nikolava D, Gluud LL, et al. Antioxidant supplements for prevention of mortality in healthy participants and patients with various diseases. Cochrane Database Syst Rev 2008 Apr; 16(2): CD007176.

21. Brewer GJ. Iron and copper toxicity in diseases of aging, particularly atherosclerosis and Alzheimer's disease. Exp Biol Med 2007; 232(2): 323.

22. Brewer GJ. Iron and copper toxicity in diseases of aging, particularly atherosclerosis and Alzheimer's disease. Exp Biol Med 2007; 232(2): 323. Morris MC, et al. Dietary copper and high saturated and trans fat intakes associated with cognitive decline. Archives of Neurology 2006; 63: 1085–8.

23. Vinceti M, Wei ET, Malagoli C, et al. Adverse health effects of selenium in humans. Rev Environ Health 2001; 16(4): 233–51. Mueller AS, Mueller K, Wolf NM, Pallauf J. Selenium and diabetes: an enigma? Free Radic Res 2009; 8:1–31. Laclaustra M, Navas-Acien A, Stranges S, et al. Serum selenium concentrations and diabetes in U.S. adults: National Health and Nutrition Examination Survey (NHANES) 2003–2004. Environ Health Perspect 2009; 117(9): 1409–13. Navas-Acien A, Bleys J, Guallar E. Selenium intake and cardiovascular risk: what is new? Curr Opin Lipidol 2008; 19(1): 43–9. Stranges S, Laclaustra M,

Ji C, et al. Higher selenium status is associated with adverse blood lipid profile in British adults. J Nutr 2010; 140(1): 81–7. Chan JM, Oh WK, Xie W, et al. Plasma selenium, manganese superoxide dismutase, and intermediate- or high-risk prostate cancer. J Clin Oncol 2009; 27(22): 3577–83.

24. Hunt JR. Bioavailability of iron, zinc, and other trace minerals from vegetarian diets. Am J Clin Nutr 2003; 78(suppl): 633S 639S. De Bortoli MC, Cozzolino SM. Zinc and selenium nutritional status in vegetarians. Biol Trace Elem Res 2009; 127(3): 228–33. Frassinetti S, Bronzetti G, Caltavuturo L, et al. The role of zinc in life: a review. J Environ Pathol Toxicol Oncol 2006; 25(3): 597–610.

25. Mitri J, Muraru MD, Pittas AG. Vitamin D and type 2 diabetes: a systematic review. Eur J Clin Nutr 2011 Sep; 65(9): 1005–15.

26. Zoler ML. High vitamin D intake linked to reduced fractures. Family Practice News 2010 (November 16, 2010). http://www.familypracticenews. com/news/diabetes-endocrinology-metabolism/single-article/high-vita-min-d-intake-linked-to-reduced-fractures/61811559c684d1757cd4985b a2c57fc5.html. Bolland MJ, Avenell A, Baron JA, et al. Effect of calcium supplements on risk of myocardial infarction and cardiovascular events: meta-analysis. BMJ 2010 Jul 29; 341: c3691. Wang L, Manson JE, Sesso HD. Calcium intake and risk of cardiovascular disease: a review of prospective studies and randomized clinical trials. Am J Cardiovasc Drugs 2012 Apr 1; 12(2): 105–16.

27. Martins JG. EPA but not DHA appears to be responsible for the efficacy of omega-3 long chain polyunsaturated fatty acid supplementation in depression: evidence from a meta-analysis of randomized controlled trials. J Am Coll Nutr 2009 Oct; 28(5): 525–42.

28. McEwen B, Morel-Kopp MC, Tofler G, Ward C. Effect of omega-3 fish oil on cardiovascular risk in diabetes. Diabetes Educ 2010 Jul–Aug; 36(4): 565–84. Von Schacky C. The omega-3 index as a risk factor for cardiovascular diseases. Prostaglandins Other Lipid Mediat 2011 Nov; 96(1–4): 94–8.

29. De Luis DA, Conde R, Aller R, et al. Effect of omega-3 fatty acids on cardiovascular risk factors in patients with type 2 diabetes mellitus and hypertriglyceridemia: an open study. Eur Rev Med Pharmacol Sci 2009 Jan–Feb; 13(1): 51–5. Hartweg J, Farmer AJ, Holman RR, Neil A. Potential impact of omega-3 treatment on cardiovascular disease in type 2 diabetes. Curr Opin Lipidol 2009 Feb; 20(1): 30–8. Hartweg J, Perera R, Montori V, et al. Omega-3 polyunsaturated fatty acids (PUFA) for type 2 diabetes mellitus. Cochrane Database Syst Rev 2008 Jan 23; (1): CD003205.

30. Djoussé L, Gaziano JM, Buring JE, Lee IM. Dietary omega-3 fatty acids and fish consumption and risk of type 2 diabetes. Am J Clin Nutr 2011 Jan; 93(1): 143–50. Brostow DP, Odegaard AO, Koh WP, et al. Omega-3 fatty acids and incident type 2 diabetes: the Singapore Chinese Health Study. Am J Clin Nutr 2011 Aug; 94(2): 520–6.

31. Kaushik M, Mozaffarian D, Spiegelman D, et al. Long-chain omega-3 fatty acids, fish intake, and the risk of type 2 diabetes mellitus. Am J Clin Nutr 2009 Sep; 90(3): 613–20.

32. Thornalley PJ, Babaei-Jadidi R, Al Ali H. High prevalence of low plasma thiamine concentration in diabetes linked to a marker of vascular disease. Diabetologia 2007 Oct; 50(10): 2164–70. Vindedzis SA, Stanton KG, Sherriff JL, Dhaliwal SS. Thiamine deficiency in diabetes—is diet relevant? Diab Vasc Dis Res 2008 Sep; 5(3): 215.

33. Page GL, Laight D, Cummings MH. Thiamine deficiency in diabetes mellitus and the impact of thiamine replacement on glucose metabolism and vascular disease. Int J Clin Pract 2011 Jun; 65(6): 684–90.

34. Thornalley PJ. The potential role of thiamine (vitamin B_1) in diabetic complications. Curr Diabetes Rev 2005 Aug; 1(3): 287–98. Page GL, Laight D, Cummings MH. Thiamine deficiency in diabetes mellitus and the impact of thiamine replacement on glucose metabolism and vascular disease. Int J Clin Pract 2011 Jun; 65(6): 684–90.

35. Arora S, Lidor A, Abularrage CJ, et al. Thiamine (vitamin B_1) improves endothelium-dependent vasodilatation in the presence of hyperglycemia. Ann Vasc Surg 2006 Sep; 20(5): 653–8. Wong CY, Qiuwaxi J, Chen H, et al. Daily intake of thiamine correlates with the circulating level of endothelial progenitor cells and the endothelial function in patients with type II diabetes. Mol Nutr Food Res 2008 Dec; 52(12): 1421–7. Vindedzis SA, Stanton KG, Sherriff JL, Dhaliwal SS. Thiamine deficiency in diabetes—is diet relevant? Diab Vasc Dis Res 2008 Sep; 5(3): 215.

36. Luong KV, Nguyen LT. The impact of thiamine treatment in the diabetes mellitus. J Clin Med Res 2012 Jun; 4(3): 153–60. Babaei-Jadidi R, Karachalias N, Ahmed N, et al. Prevention of incipient diabetic nephropathy by high-dose thiamine and benfotiamine. Diabetes 2003 Aug; 52(8): 2110–20. Rabbani N, Alam SS, Riaz S, et al. High-dose thiamine therapy for patients with type 2 diabetes and microalbuminuria: a randomised, double-blind placebo-controlled pilot study. Diabetologia 2009 Feb; 52(2): 208–12. Rabbani N, Thornalley PJ. Emerging role of thiamine therapy for prevention and treatment of early-stage diabetic nephropathy. Diabetes Obes Metab 2011 Jul; 13(7): 577–83. Stracke H, Gaus W, Achenbach U, et al. Benfotiamine in Diabetic Polyneuropathy (BENDIP): results of a randomised, double blind, placebo-controlled clinical

study. Exp Clin Endocrinol Diabetes 2008 Nov; 116(10): 600–605. Hammes HP, Du X, Edelstein D, et al. Benfotiamine blocks three major pathways of hyperglycemic damage and prevents experimental diabetic retinopathy. Nat Med 2003 Mar; 9(3): 294–9.

37. Nahas R, Moher M. Complementary and alternative medicine for the treatment of type 2 diabetes. Can Fam Physician 2009 Jun; 55(6): 591–6. Davis PA, Yokoyama W. Cinnamon intake lowers fasting blood glucose: meta-analysis. J Med Food 2011 Sep; 14(9): 884–9.

38. Kumar SN, Mani UV, Mani I. An open label study on the supplementation of gymnema sylvestre in type 2 diabetics. J Diet Suppl 2010 Sep; 7(3): 273–82. Nahas R, Moher M. Complementary and alternative medicine for the treatment of type 2 diabetes. Can Fam Physician 2009 Jun; 55(6): 591–6.

39. Nahas R, Moher M. Complementary and alternative medicine for the treatment of type 2 diabetes. Can Fam Physician 2009 Jun; 55(6): 591–6.

40. Nahas R, Moher M. Complementary and alternative medicine for the treatment of type 2 diabetes. Can Fam Physician 2009 Jun; 55(6): 591–6. Fuangchan A, Sonthisombat P, Seubnukarn T, et al. Hypoglycemic effect of bitter melon compared with metformin in newly diagnosed type 2 diabetes patients. J Ethnopharmacol 2011 Mar 24; 134(2): 422–8. Minich DM, Lerman RH, Darland G, et al. Hop and acacia phytochemicals decreased lipotoxicity in 3T3-L1 Adipocytes, Db/Db mice, and individuals with metabolic syndrome. J Nutr Metab 2010; 2010. pii: 467316. Bacardi-Gascon M, Dueñas-Mena D, Jimenez-Cruz A. Lowering effect on postprandial glycemic response of nopales added to Mexican breakfasts. Diabetes Care 2007 May; 30(5): 1264–5.

41. Weingartner O, Bohm M, Laufs U. Controversial role of plant sterol esters in the management of hypercholesterolaemia. Eur Heart J 2009 Feb; 30(4): 404–409. Federal Register. Title 21: Food and Drugs. Part 101—Food Labeling. Subpart E—Specific Requirements for Health Claims. 101.83 Health claims: plant sterol/stanol esters and risk of coronary heart disease (CHD). http://ecfr .gpoaccess.gov/cgi/t/text/text-idx?c=ecfr;sid=502078d8634923edc695b394a3 57d189;rgn=div8;view=text;node=21%3A2.0.1.1.2.5.1.14;idno=21;cc=ecfr.

42. Weingartner O, Bohm M, Laufs U. Controversial role of plant sterol esters in the management of hypercholesterolaemia. Eur Heart J 2009 Feb; 30(4): 404–409. Federal Register. Title 21: Food and Drugs. Part 101—Food Labeling. Subpart E—Specific Requirements for Health Claims. 101.83 Health claims: plant sterol/stanol esters and risk of coronary heart disease (CHD). http://ecfr .gpoaccess.gov/cgi/t/text/text-idx?c=ecfr;sid=502078d8634923edc695b394a3 57d189;rgn=div8;view=text;node=21%3A2.0.1.1.2.5.1.14;idno=21;cc=ecfr.

43. Berger A, Jones P, Abumweis S. Plant sterols: factors affecting their efficacy and safety as functional ingredients. Lipids in Health and Disease 2004 3: 5 http://www.lipidworld.com/content/3/1/5.

44. Woyengo TA, Ramprasath VR, Jones PJ. Anticancer effects of phytosterols. Eur J Clin Nutr 2009 Jul; 63(7): 813–20. Mendilaharsu M, De Stefani E, Deneo-Pellegrini H, et al. Phytosterols and risk of lung cancer: a case study in Uruguay. Lung Cancer 1998; 21: 37–45. Ronico A, De Stefani E, Boffetta P, et al. Vegetables, fruits and related nutrients and risk of breast cancer: a case control study in Uruguay. Nutr Canc 1999; 35: 111–19. De Stefani E, Brennan P, Boffeta P, et al. Vegetables, fruits, related dietary antioxidants and the risk of squamous cell carcinoma of the esophagus: a case control study in Uruguay. Nutr Canc 2000; 38: 23–29. De Stefani E, Boffetta P, Ronco AL, et al. Plant sterols and risk of stomach cancer: a case study in Uruguay. Nutr Canc 2000; 37: 140–4.

45. Aviram M, Dornfield L, Rosenblat M, et al. Pomegranate juice consumption reduces oxidative stress, atherogenic modifications to LDL, and platelet aggregation: studies in humans and in atherosclerotic apolipoprotein e-deficient mice. Am J Clin Nutr 2000; 71(5); 1062–76. Aviram M, Dornfeld L. Pomegranate juice consumption inhibits serum angiotensin converting enzyme activity and reduces systolic blood pressure. Atherosclerosis 2001 Sep; 158(1): 195–8. Aviram M, Rosenblat M, Gaitini D, et al. Pomegranate juice consumption for 3 years by patients with carotid artery stenosis reduces common carotid intima-media thickness, blood pressure and LDL oxidation. Clin Nutr 2004 Jun; 23(3): 423–33.

46. Jurenka JS. Therapeutic applications of pomegranate (Punica granatum L.): a review. Altern Med Rev 2008 Jun; 13(2): 128–44.

47. Jurenka JS. Therapeutic applications of pomegranate (Punica granatum L.): a review. Altern Med Rev 2008 Jun; 13(2): 128–44. Fenercioglu AK, Saler T, Genc E, et al. The effects of polyphenol-containing antioxidants on oxidative stress and lipid peroxidation in type 2 diabetes mellitus without complications. J Endocrinol Invest 2010 Feb; 33(2): 118–24. Aviram M, Dornfield L, Rosenblat M, et al. Pomegranate juice consumption reduces oxidative stress, atherogenic modifications to LDL, and platelet aggregation: studies in humans and in atherosclerotic apolipoprotein E-deficient mice. Am J Clin Nutr 2000; 71(5); 1062–76. Aviram M, Dornfeld L. Pomegranate juice consumption inhibits serum angiotensin converting enzyme activity and reduces systolic blood pressure. Atherosclerosis 2001 Sep; 158(1): 195–8. Aviram M, Rosenblat M, Gaitini D, et al. Pomegranate juice consumption for 3 years by patients with carotid artery stenosis reduces common carotid intima-media thickness, blood pressure and LDL oxidation. Clin Nutr 2004 Jun; 23(3): 423–33.

48. Aviram M, Rosenblat M, Gaitini D, et al. Pomegranate juice consumption for 3 years by patients with carotid artery stenosis reduces common carotid intima-media thickness, blood pressure and LDL oxidation. Clin Nutr 2004 Jun; 23(3): 423–33.

49. Balk EM, Tatsioni A, Lichtenstein AH, et al. Effect of chromium supplementation on glucose metabolism and lipids: a systematic review of randomized controlled trials. Diabetes Care 2007; 30(8): 2154–63.

Chapter 9: The Six Steps to Achieving Our Health Goals

1. Jancin B. Fitness sharply cut death in high-BMI diabetics. Family Practice News 2008; Oct 1: 19.

Chapter 10: For Doctors and Patients

1. American Diabetes Association. Economic costs of diabetes in the U.S. in 2007. Diabetes Care 2008; 31: 596–615.

2. Centers for Disease Control and Prevention. 2011 National Diabetes Fact Sheet. http://www.cdc.gov/diabetes/pubs/pdf/ndfs_2011.pdf.

3. Flegal KM, Carroll MD, Kit BK, Ogden CL. Prevalence of obesity and trends in the distribution of body mass index among U.S. adults, 1999–2010. JAMA 2012; 307(5): 491–7. Abdullah A, Stoelwinder J, Shortreed S, et al. The duration of obesity and the risk of type 2 diabetes. Public Health Nutr 2011; 14(1): 119–26. Kahn SE, Hull RL, Utzschneider KM. Mechanisms linking obesity to insulin resistance and type 2 diabetes. Nature 2006; 444(7121): 840–6.

4. Nathan DM, Buse JB, Davidson MB, et al. Medical management of hyperglycemia in type 2 diabetes: a consensus algorithm for the initiation and adjustment of therapy. Diabetes Care 2008; 32(1): 193–203.

5. Koro CE, Bowlin SJ, Bourgeois N, et al. Glycemic control from 1988 to 2000 among U.S. adults diagnosed with type 2 diabetes. Diabetes Care 2004; 27: 17–20.

6. ADVANCE Collaborative Group. Intensive blood glucose control and vascular outcomes in patients with type 2 diabetes. NEJM 2008; 358: 2560–72.

7. Fonseca V. Effect of thiazolidinediones on body weight in patients with diabetes mellitus. Am J Med 2003; 115 Suppl 8A: 42S–48S.

8. Russell-Jones D, Khan R. Insulin-associated weight gain in diabetes—causes, effects and coping strategies. Diabetes Obes Metab 2007; 9(6): 799–812.

9. Ward S, Lloyd JM, Pandor A, et al. A systematic review and economic evaluation of statins for the prevention of coronary events. Health Technol Assess 2007; 11(14): 1–178.

10. Löbner K, Knopff A, Baumgarten A, et al. Predictors of postpartum diabetes in women with gestational diabetes mellitus. *Diabetes* 2006; 55(3): 792–7.

Chapter 11: Frequently Asked Questions

1. Christakis NA, Fowler JH. The spread of obesity in a large social network over 32 years. NEJM 2007; 327(4): 370–9.

2. Obarzanek E, Sacks FM, Moore TJ. Dietary approaches to stop hypertension (DASH)—sodium trial. Paper presented at Annual Meeting of the American Society of Hypertension 2000; New York, NY.

3. Itoh R, Syuyama Y. Sodium excretion in relation to calcium and hydroxy-proline excretion in a healthy Japanese population. Am J Clin Nutr 1996; 63(5): 735–40.

4. Tuomilehto J, Jousilahti P, Rastenyte D. Urinary sodium excretion and cardiovascular mortality in Finland: a prospective study. Lancet 2001; (9259): 848–51.

5. Dallongeville J, Marecaux N, Ducmetiere P, et al. Influence of alcohol consumption and various beverages on waist girth and waist-to-hip ratio on a sample of French men and women. J Obes Relat Metab Disord 1998; 22(12): 1178–83.

6. Dumitrescu RG, Shields PG. The etiology of alcohol-induced breast cancer. Alcohol 2005; 35(3): 213–25.

7. Boyle P, Boffetta P. Alcohol consumption and breast cancer risk. Breast Cancer Res 2009; 11 Suppl 3: S3.

8. Chen WY, Rosner B, Hankinson SE, et al. Moderate alcohol consumption during adult life, drinking patterns, and breast cancer risk. JAMA 2011; 306(17): 1884–90.

9. Frost L, Vestergaard P. Alcohol and risk of atrial fibrillation or flutter: a cohort study. Arch Intern Med 2004; 164(18): 1993–98. Mukamal KJ, Tolstrup JS, Friberg J, et al. Alcohol consumption and risk of atrial fibrillation in men and women: the Copenhagen City Heart Study. Circulation 2005; 112(12): 1736–42.

10. Sanderson WT, Talaska G, Zaebst D, et al. Pesticide prioritization for a brain cancer case-control study. Environ Res 1997; 74(2): 133–144. Zahm SH, Blair A. Cancer among migrant and seasonal farmworkers: an epidemiologic review and research agenda. Am J Ind Med 1993; 24(6): 753–66.

11. Worthington V. Nutritional quality of organic versus conventional fruits, vegetables and grains. J Alt ComlMed 2001; 7(2): 161–173. Grinder-Pederson L, Rasmussen SE, Bugel S, et al. Effect of diets based on foods from conven-

tional versus organic production on intake and excretion of flavonoids and markers of antioxidative defense in humans. J Agric Food Chem 2003; 51(19): 5671–6.

12. Sari I, Baltaci Y, Bagci C, et al. Effect of pistachio diet on lipid parameters, endothelial function, inflammation, and oxidative status: a prospective study. Nutrition 2010; 26(4): 399–404.

13. Bes-Rastrollo M, Wedick NM, Martinez-Gonzalez MA, et al. Prospective study of nut consumption, long-term weight change, and obesity risk in women. Am J Clin Nutr 2009; 89(6): 1913–9. Alper CM, Mattes RD. Effects of chronic peanut consumption on energy balance and hedonics. Int J Obes Relat Metab Disord 2002; 26(8): 1129–37.

Recipe Index

Desserts

Index

A1C levels. *See* HbA1C levels

acacia extract, 150

acarbose (Precose), 180

acrylamides (toxic), 50

Action to Control Cardiovascular Risk in Diabetes study, 35

Actos (pioglitazone), 21, 32, 37, 180

adult-onset diabetes. *See* diabetes type 2

Adventist Health Study, 112, 118

AGEs (advanced glycation end products), 42

alcohol intake, 201–2

almonds: Almond Tomato Sauce, 228; ANDI score of, 49; Blueberry Cobbler, 262; protein content of, 86, 132; Russian Fig Dressing/Dip, 226; Thousand Island Dressing, 226

alpha lipoic acid, 149

Amaryl (glimpiride), 21, 37, 180

American College of Lifestyle Medicine, 189

American Diabetes Association (ADA): consensus statement (2009) issued on recommendations for type 2 diabetics, 170–71; diet approved by the, 21–23, 124; disease-causing food habits reinforced by, 36–37; on the lost war against diabetes, 9; medications promoted as accepted treatment by, 2, 36–37

American Journal of Clinical Nutrition, 117

American Journal of Ophthalmology, 32

"American Kidney Fund Warns About Impact of High-Protein Diets on Kidney Health" (Crawford), 73

amputations, 9

ANDI (Aggregate Nutrient Density Index): description of the, 46–47; Fuhrman's ANDI Scores for specific foods, 49; understanding how to use the, 47–48, 50

animal products: chicken breast, 49; comparing protein of plants and, 131–32; eggs, 77, 78; fish, 76, 77–78, 139, 246, 252; limiting consumption of, 77–78, 139; low-carbohydrate high-protein diet, 69–75, 92; milk, 49, 139; recommended nutritional amounts of, 75–76; red meats, 49, 77, 139; scallops and shrimp, 139, 252; U.S. daily consumption of, 80–81. *See also* protein

Annals of Internal Medicine, 71

antioxidants: ellagitannins, 111–12; in pomegranates, 151; processed foods low in, 50

apples: ANDI score of, 49; GI and GL of, 51; high pesticide content of, 204; Soaked Oats and Blueberries, 218

artichokes: preparing steamed, 137; Saucy Lentil Loaf, 253

artificially sweetened drinks: cola, 49, 103; nutritarian diet on, 138

arugula, 49

asparagus: ANDI score of, 9; Asparagus Polonaise, 234; low pesticide content of, 204

atherosclerotic plaque: diabetes and development of, 171–72; egg consumption and build-up of, 78; how the high-nutrient-density diet promotes regression of, 177–78; studies on how the Nrf2 protein can protect against, 129–30

Atkins diet, 50, 69, 71, 72

autoimmune reaction, 11

red meats: ground beef (85% lean), 49;
nutritarian diet disallowance of, 139;
recommended avoidance of, 77
red wine, 201–2
renal detoxification system, 62
repaglinidee (Prandin), 179, 180
resistant starch (RS): calories of food
sources of, 100; description of, 91,
95; food sources of, 97–105; health
benefits of, 95–97; prebiotic nature
of, 95. *See also* starchy foods
restaurant eating tips, 198–99
restricted glucose uptake, 27, 28
rice: black, 51; brown, 51, 86, 99, 103;
No-Meat Balls, 249; white, 51,
99, 103
Roasted Vegetable Salad with Baked
Tofu or Salmon, 252
rolled oats: Blue Apple Nut Oatmeal,
217; Blueberry Cobbler, 262; dietary
fiber in, 97; GI and GL of, 51; GL
of, 103; Quick Banana Oat Breakfast
to Go, 217; resistant starch of, 99;
Saucy Lentil Loaf, 253; Simple Bean
Burgers, 255; Soaked Oats and Blueberries, 218
romaine lettuce: ANDI score of, 49;
Black Bean Lettuce Bundles, 240;
Blended Mango Salad, 219; Green
Gorilla Blended Salad, 219. *See also*
lettuce
rosiglitazone (Avandia), 2, 21, 32, 37, 180

SAD (standard American diet): dangers
of the, 8, 9–10, 41–42, 87; excessive
use of insulin necessitated by, 15–16;
type 1 diabetes and sensitivity to, 13.
See also diet plans
salad dressings/dips: cooking techniques for making, 141–42; general
guidelines for, 135; recipes for, 220–
25; restaurant eating and, 198

salads: general guidelines for dinner,
136; general guidelines for lunch,
135. *See also specific recipes*
salmon: ANDI score of, 49; Roasted
Vegetable Salad with Baked Tofu or
Salmon, 252
salt consumption, 141, 199–201
saturated fats, 28
sauce recipes, 228–30
scallions: The Big Veggie Stir-Fry, 238;
Black Bean Lettuce Bundles, 240;
Fast Mexican Black Bean Soup, 232;
Fresh Tomato Salsa, 221; Island Black
Bean Dip, 225; as recommended
salad vegetable, 126
scallops, 139, 252
SCFAs (short-chain fatty acids), 95–96,
100
seafood: fish, 76, 77–78, 246; scallops
and shrimp, 139, 252
sea salt, 199–201
seeds: ANDI score of all, 49; best eaten
raw or lightly toasted, 120; The Big
Veggie Stir-Fry, 238; Caesar Salad
Dressing/Dip, 220; comparing without nuts/seeds and with nuts/seeds
menus, 119–20; eaten in moderation, 121; Fast Mexican Black Bean
Soup, 232; fenugreek, 150; Green
Velvet Dressing/Dip, 223; health
benefits of consumption of, 111–13;
nutritarian diet and allowed consumption of, 125, 138; pumpkin,
228; Quick Banana Oat Breakfast to
Go, 217; Russian Fig Dressing/Dip,
226; sesame, 132, 220, 222, 235, 236;
Simple Bean Burgers, 255; Soaked
Oats and Blueberries, 218; studies on
reversal of diabetes and obesity by
eating, 114–17; sunflower, 132, 223,
249. *See also* nuts
selenium, 146–47

Also Available from HarperOne

JOEL FUHRMAN, M.D.
New York Times bestselling author of EAT TO LIVE

SUPER
IMMUNITY

The Essential Nutrition Guide for Boosting Your Body's Defenses
to Live Longer, Stronger, and Disease Free

New York Times Bestseller

NO SHOTS · NO DRUGS · NO SICK DAYS

HarperOne
An Imprint of HarperCollins*Publishers*